TI-83 and TI-83 Plus Graphing Calculator Manual

for use with

Elementary Statistics
A Step by Step Approach

Fifth Edition

Allan G. Bluman
Community College of Allegheny County

Prepared by

Carolyn L. Meitler
Concordia University Wisconsin

Boston Burr Ridge, IL Dubuque, IA Madison, WI New York San Francisco St. Louis
Bangkok Bogotá Caracas Kuala Lumpur Lisbon London Madrid Mexico City
Milan Montreal New Delhi Santiago Seoul Singapore Sydney Taipei Toronto

The McGraw-Hill Companies

TI-83 and TI-83 Plus Graphing Calculator Manual for use with
ELEMENTARY STATISTICS: A STEP BY STEP APPROACH, FIFTH EDITION
ALLAN G. BLUMAN

Published by McGraw-Hill Higher Education, an imprint of The McGraw-Hill Companies, Inc., 1221 Avenue of the Americas, New York, NY 10020.

1 2 3 4 5 6 7 8 9 0 DCD/DCD 0 9 8 7 6 5 4 3

ISBN 0-07-254916-5

www.mhhe.com

CONTENTS

PREFACE

This manual is written to help you use the power of the Texas Instruments ® TI-83 and TI-83+ graphing calculators to learn about statistics and to solve exercises found in Bluman's *Elementary Statistics: A Step by Step Approach*, Fifth Edition.

Examples

Examples show how a graphing calculator can be used to solve many types of statistical problems. Detailed solutions are included that illustrate all the keystrokes necessary to solve the problems. Several examples show more than one method of solution using the calculator. No parametric equations or programming on the calculators is used.

Exercise sets

Exercise sets are found at the end of each chapter in the manual. These problems are similar to the examples in the textbook, and provide an opportunity to practice the techniques demonstrated. Solutions to selected exercises are found at the end of this manual.

Calculations

All calculations in this manual have been performed using the default settings on the calculator unless otherwise noted. Numbers are rounded to the desired number of decimal places after all calculations have been completed.

❖Also available for student purchase to improve your success in this course when using *Elementary Statistics: A Step by Step Approach,* Fifth Edition by Allan G. Bluman:

Student Study Guide
By Pat Foard of South Plains College, this study guide will assist students in understanding and reviewing key concepts and preparing for exams. It emphasizes all important concepts contained in each chapter, includes explanations, and provides opportunities for students to test their understanding by completing related exercises and problems.

Student Solutions Manual
By Sally Robinson of South Plains College, this manual contains detailed solutions to all odd-numbered text problems and answers to all quiz questions.

Critical Thinking Workbook
By James Condor of Manatee Community College, this workbook provides a number of additional challenging problems for students to solve that are drawn from real-world applications. Problems are tied to each chapter of the textbook and highlight and reinforce key concepts.

Excel Manual for Office 2000
By Renee Goffinet and Virginia Koehler of Spokane Falls Community College, this workbook is specially designed to accompany the textbook and provides additional practice in applying the chapter concepts while using Excel.

MINITAB Manual
By Gerry Moultine of Northwood University, this manual provides the student with how-to information on data and file management, conducting various statistical analyses, and creating presentation-style graphics while following each text chapter.

Videos
New to this edition are text-specific videos available on VHS and CD-ROM that demonstrate key concepts and worked-out exercises from the text plus tutorials in using the TI-83 Plus Calculator, Excel, and MINITAB, in a dynamic, engaging format.

ACKNOWLEDGMENTS

The completion of this project would not have occurred without the support and assistance of several individuals. Specifically, I would thank my family for their encouragement and support.

Carolyn L. Meitler

CHAPTER 1

BASIC TI-83 AND TI-83+ OPERATIONS

Getting Started

	TI-83	TI-83+
Press [ON] to turn on the calculator. Press [2nd] [+] to get the MEMORY screen. Use the down arrow [▼] to select 5:Reset... on the TI-83 or 7:Reset… on the TI-83+. Press [ENTER].	MEMORY 1:Check RAM... 2:Delete... 3:Clear Entries 4:ClrAllLists 5:Reset...	MEMORY 1:About 2:Mem Mgmt/Del... 3:Clear Entries 4:ClrAllLists 5:Archive 6:UnArchive 7↓Reset...
On the TI-83 select 1:All Memory... and press [ENTER].	RESET 1:All Memory... 2:Defaults...	RAM ARCHIVE ALL 1:All RAM... 2:Defaults...
On the TI-83+ use the right arrow to select 1:All Memory... and press [ENTER].		RAM ARCHIVE ALL 1:All Memory...
Use the down arrow [▼] to choose 2:Reset and press [ENTER].	RESET MEMORY 1:No 2:Reset Resetting memory erases all data and programs.	RESET MEMORY 1:No 2:Reset Resetting ALL will delete all data, programs & Apps from RAM & Archive.
The screen should now indicate that the memory is cleared. However, the screen may look blank. This is because the contrast setting may also have been reset and now needs to be adjusted.	Mem cleared	TI-83 Plus 1.03 Mem cleared

Press [2nd] and then hold the [▲] key depressed until you see the display in the middle of the screen. Now the contrast will be dark enough for you to see the screen display.

If the contrast is too dark, press [2nd] and hold the [▲] depressed until the screen is the contrast you want.

> Press 2nd [▼] to make the display darker.
>
> Press 2nd [▲] to make the display lighter.

To check the battery power, press 2nd [▲] and note the number that appears in the upper right corner of the screen. If it is an 8 or 9, you should replace your batteries. The lowest number is 0 and the highest number is 9.

> Press [CLEAR] to clear the screen.
>
> Press [2nd][OFF] to turn off the calculator.

Special Keys, Home Screen and Menus

[2nd]

This key must be pressed to access the operation above and to the left of a key. These operations are a yellow color on the face of the calculator. When you press the [2nd] key a flashing up arrow ↑ is displayed on the screen.

> In this document, the functions on the face of the calculator above a key will be referred to in boxes just as if the function was printed on the key cap. For example, [ANS] is the function above the [(-)] key.

[ALPHA]

This key must be pressed first to access the operation above and to the right of a key. A flashing [A] is displayed as the cursor on the screen after the [ALPHA] key is pressed.

[A-LOCK]

[2nd][A-LOCK] locks the calculator into alpha mode. The calculator will remain in alpha mode until the [ALPHA] is pressed again.

[MODE]

Press [MODE]. The highlighted items are active. Select the item you wish using the arrow keys. Press [ENTER] to activate the selection.

Normal Sci Eng	Type of notation for display of numbers.
Float 0123456789	Number of decimal places displayed.
Radian Degree	Type of angle measure.
Func Par Pol Seq	Function or parametric graphing.
Connected Dot	Connected or not connected plotted points on graphs.
Sequential Simul	Graphs functions separately or all at once.
Real a+bi re^θi	Allows number to be entered in rectangular complex mode or polar complex mode.
Full Horiz G-T	Allows a full screen or split screen to be used.

```
Normal Sci Eng
Float 0123456789
Radian Degree
Func Par Pol Seq
Connected Dot
Sequential Simul
Real a+bi re^θi
Full Horiz G-T
```

Home Screen

The screen on which calculations are done and commands are entered is called the Home Screen. You can always get to this screen (aborting any calculations in progress) by pressing
QUIT
[2nd] [MODE] . From here on, this will be referred to as [2nd] [QUIT] in this manual.

Menus

The TI-83+ Graphics calculator uses menus for selection of specific functions. The items on the menus are identified by numbers followed by a colon. There are two ways to choose menu items:

1. Using the arrow keys to highlight the selection and then pressing [ENTER] .
2. Pressing the number corresponding to the menu item.

In this document the menu items will be referred to using the key to be pressed followed by the meaning of the menu. For example, on the [ZOOM] menu, [1] :ZBox refers to the first menu item.

Correcting Errors

It is easy to correct errors on the screen when entering data into the calculator. To do so use the arrow keys, [DEL] , and [INS] keys.

[◄] or [►]	Moves the cursor to the left or right one position.
[▼] or [►]	Moves the cursor up or down one line.
[DEL]	Deletes one or more characters at the cursor position.
[2nd] [INS]	Inserts one or more characters at the cursor position.

Calculation

Example 1 Calculate $\dfrac{8.5 - 8.9}{\dfrac{1.8}{\sqrt{31}}}$.

Turn the calculator on and press [2nd] [QUIT] to return to the Home Screen.

Press [CLEAR] to clear the Home Screen. Now we are ready to do a new calculation.

Numbers and characters are entered in the same order as you would read an expression. Do not press [ENTER] unless specifically instructed to do so in these examples. Keystrokes are written in a column but you should enter all the keystrokes without pressing the [ENTER] key until [ENTER] is displayed in the example.

Solution:

Keystrokes	*Screen Display*	*Comments*
[2nd] [QUIT] [CLEAR]		It is a good idea to clear the screen before starting a calculation.
[(] [8.5] [-] [8.9] [)] [÷] [(] [1.8] [÷] [2nd] [√] [31] [)] [)] [ENTER]	(8.5-8.9)/(1.8/√(31)) -1.23728097	The lines on the calculator will wrap around. Do not press [ENTER] until you have finished entering all numbers and symbols. The answer is -1.24 rounded to two decimal places

Notice that the negative sign in the answer is different from the subtraction sign in the numerator.
Notice, also, that a left parentheses is entered automatically with the square root operation.

Setting the WINDOW and Graphing

Before doing any graphing on the calculator, the statistical graphing commands need to be turned off.

[2nd] [STAT PLOT] [4] :PlotsOff [ENTER]

Also, drawings need to be cleared.

[2nd] [DRAW] [1] :ClrDraw [ENTER]

Furthermore, all functions should be cleared from the function list.

[Y=] [CLEAR] [ENTER] [CLEAR] etc.

Example 1

Graph $y = 3.4 + 1.3x$, $y = 3.4 + 1.3 \ln x$, $y = 3.4(1.3^x)$ and $y = 3.4x^{1.3}$ on the same coordinate axes.

(A) Graph the first function with a thin line, the second function with a bold line, the third function with a dotted line, and the fourth function with a thin line. Graph using the standard graph screen dimensions of $-10 \le x \le 10$ and $-10 \le y \le 10$ with a scale of 1 on the x axis and a scale of 1 on the y axis.

(B) Change the graph screen dimensions to $-5 \le x \le 5$ and $-10 \le y \le 20$, with a scale of 1 on the x axis and a scale of 2 on the y axis. Deselect the first and fourth functions so they do not graph but remain in the [Y=] list so they can be evaluated.

Solution (A):

Keystrokes	*Screen Display*	*Comments*
[2nd] [STAT PLOT]	STAT PLOTS 1:Plot1…Off L1 L2 2:Plot2…Off L1 L2 3:Plot3…Off L1 L2 4↓PlotsOff	Get the STAT PLOT menu.
[4] :PlotsOff [ENTER]	PlotsOff Done	Turn all statistical plots off
[2nd] [DRAW]	DRAW POINTS STO 1:ClrDraw 2:Line(3:Horizontal 4:Vertical 5:Tangent(6:DrawF 7↓Shade(	Get the DRAW menu.
[1] :ClrDraw [ENTER] [CLEAR] [3.4] [+] [1.3] [X,T,θ,n] [ENTER]	PlotsOff Done ClrDraw Done	Clear all drawings. Clear the existing Y1 function and store the first function as Y1.
[CLEAR] [3.4] [+] [1.3] [LN] [X,T,θ,n] [)] [ENTER]	Plot1 Plot2 Plot3 \Y1=3.4+1.3X \Y2=3.4+1.3ln(X) \Y3= \Y4= \Y5= \Y6=	Clear and store the second function as Y2. A right parentheses needs to be entered since the calculator automatically puts the left one in.
[CLEAR] [3.4] [(] [1.3] [^] [X,T,θ,n] [)] [ENTER]		Clear and store the third function as Y3.
[CLEAR] [3.4] [X,T,θ,n] [^] [1.3] [ENTER]	Plot1 Plot2 Plot3 \Y1=3.4+1.3X \Y2=3.4+1.3ln(X) \Y3=3.4(1.3^X) \Y4=3.4X^1.3 \Y5= \Y6=	Clear and store the fourth function as Y4.
[▲] [▲] [▲] [◄] [◄] [ENTER] ...[ENTER]	Plot1 Plot2 Plot3 \Y1=3.4+1.3X \Y2=3.4+1.3ln(X) \Y3=3.4(1.3^X) \Y4=3.4X^1.3 \Y5= \Y6=	Go to the symbol to the left of Y2. Press [ENTER] repeatedly until the bold line appears.

▼ ENTER ...ENTER	Plot1 Plot2 Plot3 \Y1■3.4+1.3X \Y2■3.4+1.31n(X) ·.Y3■3.4(1.3^X) \Y4■3.4X^1.3 \Y5= \Y6=	Press the down arrow and repeatedly press enter to change the symbol to the left of Y3 to a dotted line. Change Y1 and Y4 if they are not thin lines (the default setting).
ZOOM 6 :ZStandard		Choose the ZStandard option from the ZOOM menu. Note the ZStandard option automatically sets the graph screen dimentions at $-10 \leq x \leq 10$ and $-10 \leq y \leq 10$.

Solution (B):

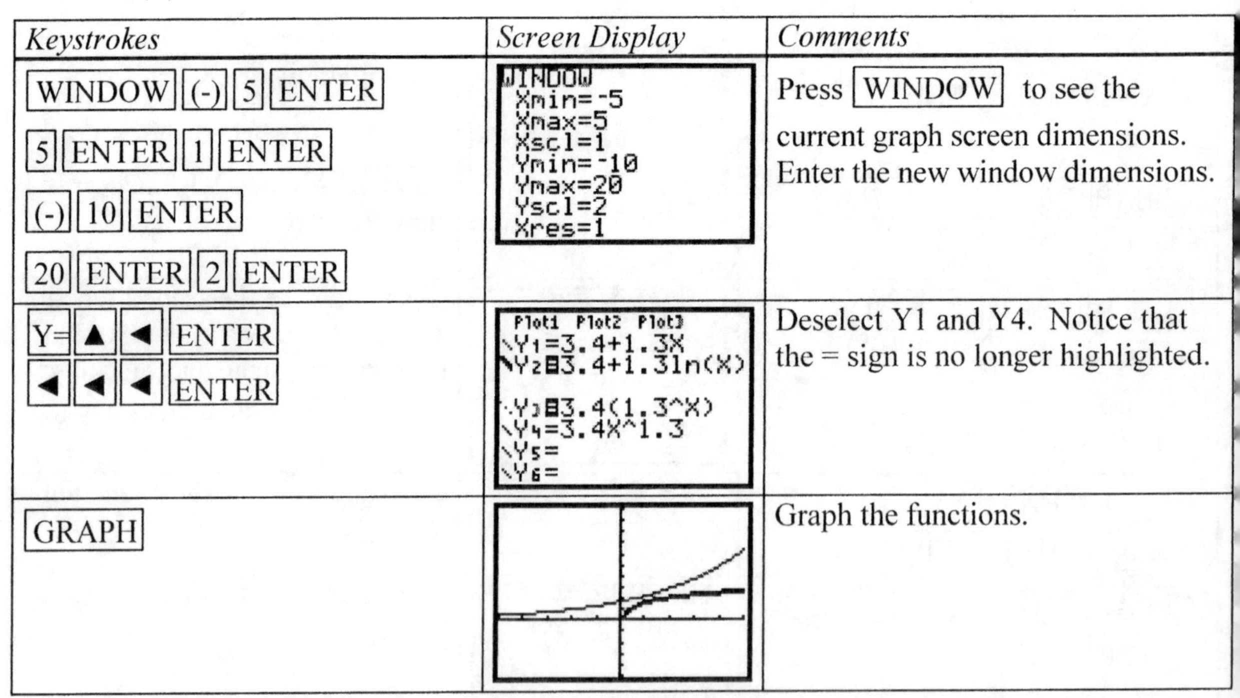

Keystrokes	*Screen Display*	*Comments*
WINDOW (-) 5 ENTER 5 ENTER 1 ENTER (-) 10 ENTER 20 ENTER 2 ENTER	WINDOW Xmin=-5 Xmax=5 Xscl=1 Ymin=-10 Ymax=20 Yscl=2 Xres=1	Press WINDOW to see the current graph screen dimensions. Enter the new window dimensions.
Y= ▲ ◄ ENTER ◄ ◄ ◄ ENTER	Plot1 Plot2 Plot3 \Y1=3.4+1.3X \Y2■3.4+1.31n(X) ·.Y3■3.4(1.3^X) \Y4=3.4X^1.3 \Y5= \Y6=	Deselect Y1 and Y4. Notice that the = sign is no longer highlighted.
GRAPH		Graph the functions.

> You may see a moving bar in the upper right corner of the screen. This means that calculator is working. Wait a few seconds for it to finish working before entering new information.

> The ZStandard screen automatically sets the graph for $-10 \leq x \leq 10$ and $-10 \leq y \leq 10$. Press ZOOM 6 :ZStandard WINDOW to see this.

> The graphs will be plotted in order: Y1, then Y2, then Y3, then Y4, etc.

TRACE, ZOOM and WINDOW

[TRACE] allows you to observe both the x and y coordinate of a point on the graph as the cursor moves along the graph. If there is more than one function graphed, the up [▲] and down [▼] arrow keys allow you to move between the graphs displayed.

[ZOOM] will magnify a graph so the coordinates of a point can be approximated with greater accuracy. Three methods to zoom in are:

1. Change the graph screen dimensions using the [WINDOW] key.
2. Use the [2] :Zoom In option on the [ZOOM] menu in conjunction with [ZOOM] [►] [4] :Set Factors.
3. Use the [1] :Zbox option on the [ZOOM] menu.
4. Use the SOLVER feature of the calculator.
5. Use the INTERSECT feature of the calculator.

Example 1 Approximate to one decimal place the value of x for $y = 5.0$ for $y = 3.4 \times 1.3^x$ using Methods 1 to 5 listed above. Round your answer to two decimal places.

Solution:

Graph the functions using the standard graphing window. (See preceding section.)

Method 1 Change the WINDOW values

Keystrokes	*Screen Display*	*Comments*
[Y=] [CLEAR] [ENTER] [CLEAR] [ENTER] [CLEAR] [3.4] [(] [1.3] [^] [X,T,θ,*n*] [)] [ENTER] [CLEAR] [ENTER] [CLEAR] [ENTER] [CLEAR] [ENTER] [ZOOM] [6] :Zstandard		Clear all functions. Clear all other functions. Enter the function as Y3. See Example 1. Graph using the standard graph screen dimensions.
[TRACE]	Y3=3.4(1.3^X) X=0 Y=3.4	[TRACE] places a cursor at $x = 0$. Note here that the y coordinate is 3.4.

▶ …▶	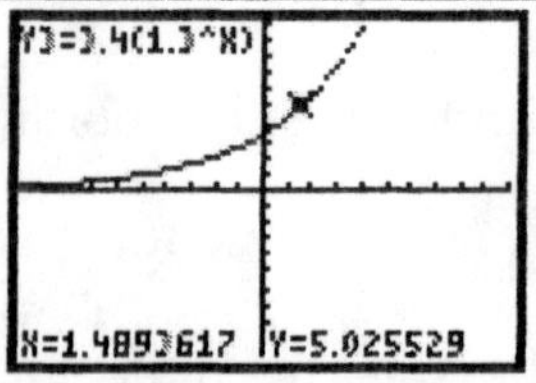	Use the right arrow to place the cursor close to $y=5$ at (1.49, 5.03). [Note: coordinates have been rounded to two decimal places.] A first approximation for the x coordinate is 1.49.
WINDOW 1.4 ENTER 1.6 ENTER 1 ENTER 4 ENTER 6 ENTER 1 ENTER	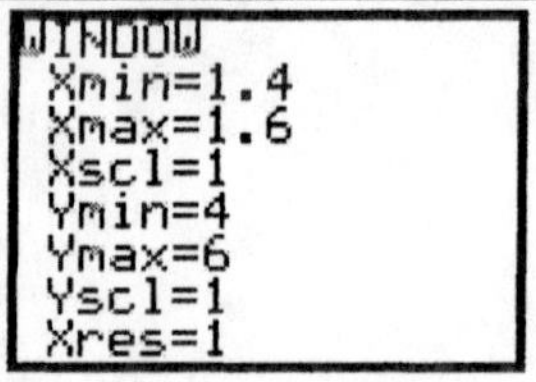 WINDOW Xmin=1.4 Xmax=1.6 Xscl=1 Ymin=4 Ymax=6 Yscl=1 Xres=1	Change the graph screen dimensions to $1.4 \le x \le 1.6$ and $4 \le y \le 6$ to get a better approximation for x.
GRAPH	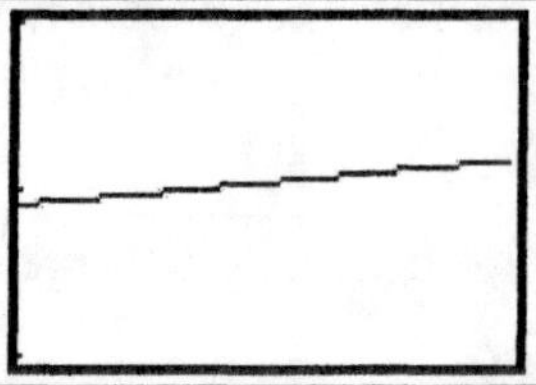	Graph the function.
TRACE	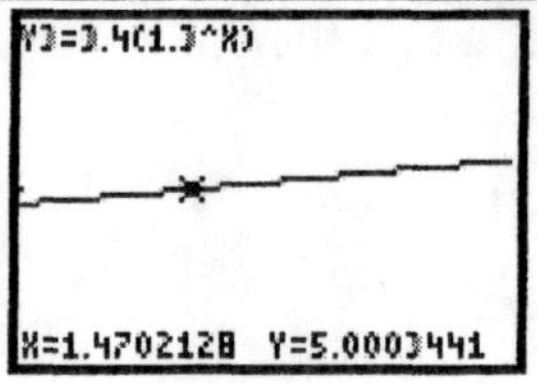	Use TRACE to get a better approximation for x. By moving the cursor to the right and left to a value closest to $y=5.0$, we see that the value of the x coordinate, rounded to two decimal place, is 1.50.

Method 2 Use Zoom In

Keystrokes	*Screen Display*	*Comments*
Y= CLEAR ▼ ▼ CLEAR 3.4 (1.3 ^ X,T,θ,n) ENTER CLEAR ENTER CLEAR ENTER CLEAR ENTER ZOOM 6 :ZStandard ENTER	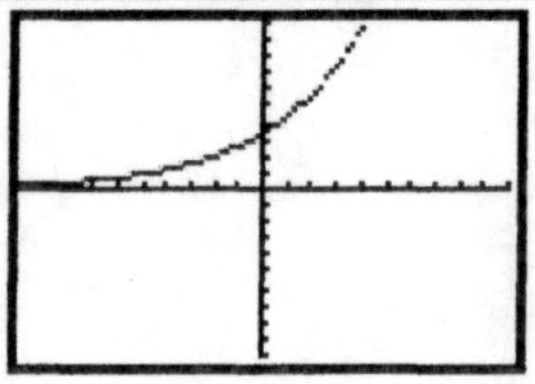	Clear all functions. Enter the function as Y3. See Example 1. Clear all other functions. Graph using the standard graph screen dimensions.

ZOOM ▶	ZOOM MEMORY 1:ZPrevious 2:ZoomSto 3:ZoomRcl 4:SetFactors…	Get the MEMORY menu.
4 :Set Factors 4 ENTER 4	ZOOM FACTORS XFact=4 YFact=4	Set the zoom factors to 4 and 4.
ZOOM 2 :Zoom In ▶ …▲	X=1.4893617 Y=5.1612903	Use zoom in and place the cursor as shown. The cursor is a + sign.
ENTER	X=1.4893617 Y=5.1612903	The new graphing screen will be centered at the point the + sign was positioned. Use TRACE and the arrow keys again to get a better approximation for the *x* coordinate.
TRACE	Y3=3.4(1.3^X) X=1.4893617 Y=5.025529	By moving the cursor to the right and left, we see that the value of the *x* coordinate, rounded to two decimal place, is 1.50.

Method 3 Use ZBox

Keystrokes	*Screen Display*	*Comments*
Y= CLEAR ENTER CLEAR ENTER CLEAR 3.4 (1.3 ^ X,T,θ,*n*) ENTER CLEAR ENTER CLEAR ENTER CLEAR ENTER ZOOM 6 :Zstandard ENTER		Clear and enter the function as Y3. See Example 1. Clear all other functions. Graph using the standard graph screen dimensions.
ZOOM	ZOOM MEMORY 1:ZBox 2:Zoom In 3:Zoom Out 4:ZDecimal 5:ZSquare 6:ZStandard 7↓ZTrig	Get the ZOOM menu.

[1] :ZBox [▶] ...[▲] [ENTER]	X=.85106383 Y=5.8064516	Press 1 to get the ZBox option. Use the arrow keys to place the cursor for the upper left corner of the box at (0.85, 5.81), rounded to two decimal places. Press [ENTER] to adjust the corners of the zoom box.
[▶] ...[▼]	X=2.3404255 Y=4.516129	Use the arrow keys to place the lower right corner of the box at a point below and to the right of the curve.
[ENTER]	X=1.5957447 Y=5.1612903	Press enter to zoom in. The region inside the box is the new graphing screen.
[TRACE] [▶] ...[▶]	Y3=3.4(1.3^X) X=1.4848348 Y=5.0195637	Use [TRACE] and the arrow keys again to get a better approximation for the x coordinate. By moving the cursor to the right and left, we see that the value of the x coordinate, rounded to one decimal place, is 1.5.

[TRACE] enables the * cursor to travel along the curve displaying the coordinates of points on the curve. Using the arrow keys without pressing [TRACE] first causes the + cursor to move to any point on the screen.

Method 4 Using Solver

Keystrokes	*Screen Display*	*Comments*
[Y=] [CLEAR] [ENTER] [CLEAR] [ENTER] [CLEAR] [3.4] [(] [1.3] [^] [X,T,θ,*n*] [)] [ENTER] [CLEAR] [ENTER] [CLEAR] [ENTER] [CLEAR] [ENTER]		Clear and enter the function as Y3. See Example 1. Clear all other functions.

ZOOM 6 :Zstandard		Graph using the standard graph screen dimensions
MATH ▼ ▼ ... ▼ 0 :Solver CLEAR VARS ► :Y-Vars 1 :Function... 3 :Y3 - 5 ENTER ALPHA SOLVE	MATH NUM CPX PRB 1:►Frac 2:►Dec 3:³ 4:³√(5:ˣ√ 6:fMin(7↓fMax(	Get the solver function from the MATH menu. Enter the expression Y3-5. We wish to have this expression equal to 0 which is the same as having $3.4(1.3^x) = 5$. The result rounded to two decimal places is 1.50.

Method 5 Using Intersect

Keystrokes	*Screen Display*	*Comments*
Y= CLEAR ENTER CLEAR ENTER CLEAR 3.4 (1.3 ^ X,T,θ,*n*) ENTER CLEAR 5 ENTER CLEAR ENTER		Clear and enter the function as Y3. Enter 5 as Y4. See Example 1. Clear all other functions.
ZOOM 6 :Zstandard		Graph using the standard graph screen dimensions
2nd CALC 5 :Intersect ENTER ENTER ► ... ► ENTER		Click on the first curve and on the second curve. Move the cursor close to the intersection point and press enter. Pressing enter again will calculate the intersection point at (1.50, 5.00).

EXERCISE SET 1

1. Graph the four functions on the same set of coordinate axes:

$y = 2.1 - 1.9x$
$y = 2.1 - 1.9 \ln x$
$y = 2.1(1.9^x)$
$y = 2.1x^{1.9}$

Graph the first using a bold line, the second with a thin line, the third with a dotted line, and the fourth with a thin line.
(A) Use the standard graphing screen ($-10 \le x \le 10$ and $-10 \le y \le 10$).
(B) Use graph screen dimensions of $-5 \le x \le 5$ and $-10 \le y \le 20$ with x axis scale of 1 and y axis scale of 2.

2. Graph the four functions on the same set of coordinate axes:

$y = 5.9 - 3.2x$
$y = 5.9 - 3.2 \ln x$
$y = 5.9(3.2^x)$
$y = 5.9x^{3.2}$

(A) Use the standard graphing screen.
(B) Use graph screen dimensions of $-8 \le x \le 8$ and $-20 \le y \le 20$ with x axis scale of 1 and y axis scale of 2.

3. Approximate, to two decimal places, the x coordinate for the point having y coordinate of 4 using graphical methods for $y = 3.5x^{1.7}$. Solve using all five methods demonstrated above.

4. Approximate, to two decimal places, the x coordinate for the point having y coordinate of 3 using graphical methods for $y = 2.8(1.2^x)$. Solve using all five methods demonstrated above.

CHAPTER 2

FREQUENCY DISTRIBUTIONS AND GRAPHS

Example 1

The number of games won by the pitchers who were inducted into the Baseball Hall of Fame through 1992 are shown below.

(A) Construct a histogram with lowest class boundary of 20, highest class boundary of 420, and class width of 20.

(B) Construct a histogram with lowest class boundary of 15.5, highest class boundary of 435.5, and class width of 30.

(C) Discuss how the histograms differ.

Games Won	*Games Won*	*Games Won*	*Games Won*	*Games Won*
373	254	237	243	308
210	266	253	201	266
239	114	224	373	286
329	236	284	247	273
198	361	416	207	243
326	251	160	360	311
215	189	344	268	363
21	270	165	240	48
150	300	207	314	197
209	210	260	327	

Solution (A):

Keystrokes	*Screen Display*	*Comments*
Y= CLEAR ▼ CLEAR …	Plot1 Plot2 Plot3 \Y1= \Y2= \Y3= \Y4= \Y5= \Y6= \Y7=	Clear all functions from the calculator.

2nd STAT PLOT 4 :PlotsOff ENTER	STAT PLOTS 1:Plot1…Off L1 1 2:Plot2…Off L1 L2 3:Plot3…Off L1 L2 4↓PlotsOff PlotsOff Done	Turn all statistical plots off.
2nd DRAW 1 :ClrDraw ENTER	DRAW POINTS STO 1:ClrDraw 2:Line(3:Horizontal 4:Vertical 5:Tangent(6:DrawF 7↓Shade(PlotsOff Done ClrDraw Done	Clear all drawing from the calculator.
WINDOW 20 ENTER 420 ENTER 20 ENTER 0 ENTER 10 ENTER 1 ENTER	WINDOW Xmin=20 Xmax=420 Xscl=20 Ymin=0 Ymax=10 Yscl=1 Xres=1	Set the graph screen parameters at Xmin=20, Xmax=420, Xscl=20 (class width), Ymin=0, Ymax=10, Yscl=1, Xres=1. The Xscl is the class width. The Ymin, Ymax, and Yscl are the least frequency, greatest frequency, and scale marks for the frequency. Since we do not know what the greatest frequency is, pick a number to try. We used 10.
STAT 4 :ClrList 2nd L1 ENTER	EDIT CALC TESTS 1:Edit… 2:SortA(3:SortD(4:ClrList 5:SetUpEditor PlotsOff Done ClrDraw Done ClrList L1 Done	Clear L1.

STAT 1 :Edit... ENTER 373 ENTER 210 ENTER 239 ENTER 329 ENTER etc.	EDIT CALC TESTS 1:Edit... 2:SortA(3:SortD(4:ClrList 5:SetUpEditor L1 L2 L3 1 ------ ------ L1(1) = L1 L2 L3 1 273 243 311 363 48 197 L1(50) =	Enter data. If you make an error, use the arrow keys to highlight the entry. Reenter the number. If you need to delete an entry, use the arrow keys to highlight the entry. Press DEL .
2nd STAT PLOT 1 :Plot1 ENTER	STAT PLOTS 1:Plot1...Off L1 1 2:Plot2...Off L1 L2 3:Plot3...Off L1 L2 4↓PlotsOff	Turn Plot 1 On.
▼ ► ► ENTER ▼ 2nd L1 ENTER ALPHA 1	Plot1 Plot2 Plot3 On Off Type: Xlist:L1 Freq:1	Select the Histogram symbol and press enter. Select L1. Frequencies are 1. If you see an A flashing, change it to a square box flashing by pressing ALPHA
GRAPH		Get the graph.

Solution (B):

Keystrokes	*Screen Display*	*Comments*
[WINDOW] [15.5] [ENTER] [435.5] [ENTER] [30] [ENTER] [0] [ENTER] [10] [ENTER] [1] [ENTER] [GRAPH]	WINDOW Xmin=15.5 Xmax=435.5 Xscl=30 Ymin=0 Ymax=10 Yscl=1 Xres=1	Change the graph screen dimensions and graph again.

Solution (C):

The graph in part (A) has narrower bars than the graph in part (B) because the class width is smaller in part (A). Also, the frequency in some of the classes in Part (B) is greater since there are more numbers in some of the classes.

Example 2

Use the data in Example 1. Add the data value 20. Create a histogram and frequency polygon for this data using 20 as the lowest class boundary and 420 as the highest class boundary with a class width of 40. Discuss the relationship of the frequency polygon to the histogram.

Solution:

There are two ways to count the number of pieces of data in each class:

1. Calculate the class boundries, sort the data and count the number of pieces of data in each class by hand.
2. Get the historgram and use [TRACE] while the histogram is displayed.

Method 1 Sort Data

Keystrokes	*Screen Display*	*Comments*
[CLEAR] [STAT] [2] :SortA([2nd] [L1] [)] [ENTER]	SortA(L1) Done	Enter the data as was done on the bottom of page 14 and the tope of page 15 of this manual. Add the value 20 to the list. Sort the data using the calculator to help construct the frequency distribution.

[STAT] [1] :Edit	L1 L2 L3 1 20 21 48 114 150 160 165 L1(1)=20	Count the number of pieces of data in each class. See frequency distribution following Method 2 following.

Method 2 Use the Histogram

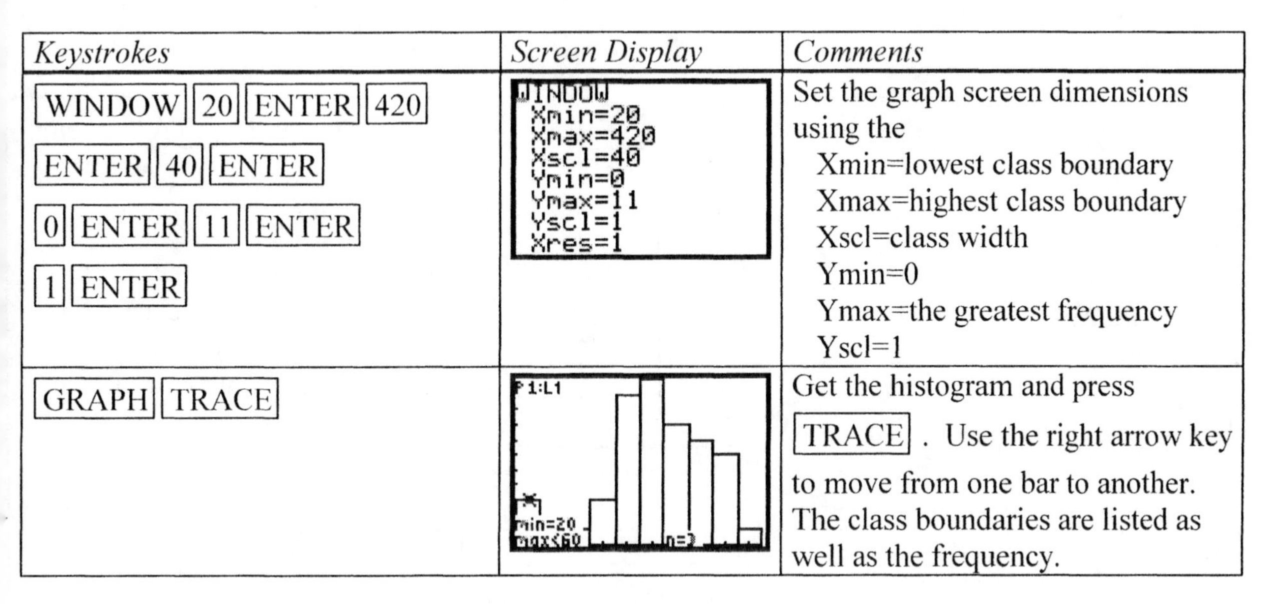

Keystrokes	*Screen Display*	*Comments*
[WINDOW] [20] [ENTER] [420] [ENTER] [40] [ENTER] [0] [ENTER] [11] [ENTER] [1] [ENTER]	WINDOW Xmin=20 Xmax=420 Xscl=40 Ymin=0 Ymax=11 Yscl=1 Xres=1	Set the graph screen dimensions using the Xmin=lowest class boundary Xmax=highest class boundary Xscl=class width Ymin=0 Ymax=the greatest frequency Yscl=1
[GRAPH] [TRACE]	P1:L1 min=20 max<60 n=3	Get the histogram and press [TRACE] . Use the right arrow key to move from one bar to another. The class boundaries are listed as well as the frequency.

We need to add an additional line (additional class) at the beginning and the end of the frequency distribution in order to construct a frequency polygon. The frequency distribution is:

Games Won	Class Midpoints x	Number of Pitchers (Frequencies) y
less than 20	0	0
at least 20 and less than 60	40	3
at least 60 and less than 100	80	0
at least 100 and less than 140	120	1
at least 140 and less than 180	160	3
at least 180 and less than 220	200	10
at least 220 and less than 260	240	11
at least 260 and less than 300	280	8
at least 300 and less than 340	320	7
at least 340 and less than 380	360	6
at least 380 and less than 420	400	1
at least 420	440	0

Note that additional points were added to the frequency distribution so that the line on the frequency polygon will begin and end on the horizontal axis at 0.

Keystrokes	*Screen Display*	*Comments*
[WINDOW] [(-)] [20] [ENTER] [460] [ENTER] [40] [ENTER] [0] [ENTER] [11] [ENTER] [1] [ENTER]	WINDOW Xmin=-20 Xmax=460 Xscl=40 Ymin=0 Ymax=11 Yscl=1 Xres=1	We wish to graph the frequency polygon and the histogram at the same time. We need to set the graph screen dimensions to Xmin= -20, Xmax=460, Xscl=40, Ymin=0, Ymax=11, Yscl=1.
[STAT] [ENTER] [▶] [▲] [CLEAR] [▼] [▶] [▲] [CLEAR] [▼]	L1: 20, 21, 48, 114, 150, 160, 165 L2 = L1: 20, 21, 48, 114, 150, 160, 165 L2(1)=	Get [EDIT] on the [STAT] menu. Use the arrow keys to move to L2. Use the arrow keys to move up, press [CLEAR] , and move down. This is another way to clear a list. Repeat for L3 so that both lists are clear.
[0] [ENTER] [40] [ENTER] [80] [ENTER] [120] [ENTER] etc. [▶] [0] [ENTER] [3] [ENTER] [0] [ENTER] [1] [ENTER] etc.	L1: 20, 21, 48, 114, 150, 160, 165 L2: 0, 40, 80, 120, 160, 200, 240 L3: 0, 3, 0, 1, 3, 10, 11 L2(1)=0 L1: 165, 189, 197, 198, 201, 207, 207 L2: 240, 280, 320, 360, 400, 440 L3: 11, 8, 7, 6, 1, 0 L2(13) =	Enter the midpoints of the classes in L2. Enter the frequencies in L3. Note that additional points were added to the frequency distribution so that the line on the frequency polygon will begin and end on the horizontal axis at 0.
[2nd] [STAT PLOT] [▼] [ENTER] [ENTER] [▼] [▶] [ENTER] [▼] [▼] [2] [L2] [▼] [2nd] [L3] [ENTER] [ENTER]	Plot1 Plot2 Plot3 On Off Type: Xlist:L2 Ylist:L3 Mark: ▫ + ·	Turn Plot2 on. Select the line graph. Select L2 as the Xlist and L3 as the Ylist.
[ZOOM] 9 :Zoom Stat		Graph using 9 :Zoom Stat or set the WINDOW at $0 \le x \le 440$ with a scale of 40 and $0 \le y \le 11$ with a scale of 1.

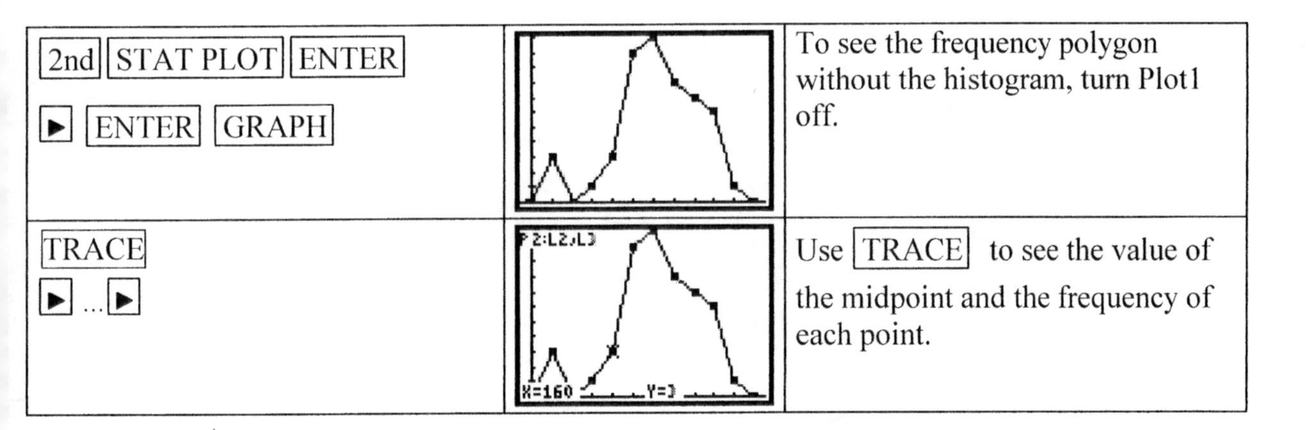

Keystrokes	Screen Display	Comments
2nd STAT PLOT ENTER ▶ ENTER GRAPH		To see the frequency polygon without the histogram, turn Plot1 off.
TRACE ▶ … ▶	P2:L2,L3 X=160 Y=3	Use TRACE to see the value of the midpoint and the frequency of each point.

Example 3

Use the data in Example 2 above. Create an ogive for this data. Use 20 as the lowest class boundary and a class width of 40.

Solution:

Extend the frequency distribution in Example 2 to include a column of cumulative frequencies. Add a row at the beginning so the cumulative frequency column begins at 0.

Number of Games Won	Number of Pitchers (Frequencies) y	Number of Games Won (Lower Class Boundaries) x	Cumulative Frequency y
		less than 20	0
at least 20 and less than 60	3	less than 60	3
at least 60 and less than 100	0	less than 100	3
at least 100 and less than 140	1	less than 140	4
at least 140 and less than 180	3	less than 180	7
at least 180 and less than 220	10	less than 220	17
at least 220 and less than 260	11	less than 260	28
at least 260 and less than 300	8	less than 300	36
at least 300 and less than 340	7	less than 340	43
at least 340 and less than 380	6	less than 380	49
at least 380 and less than 420	1	less than 420	50

Keystrokes	*Screen Display*	*Comments*
Y= CLEAR etc.	Plot1 Plot2 Plot3 \Y1= \Y2= \Y3= \Y4= \Y5= \Y6= \Y7=	Clear all functions from the Y= list.

Keystrokes	Screen	Explanation
STAT 4 :ClrList 2nd L1 , 2nd L2 , 2nd L3 ENTER 2nd STAT PLOT 4 :PlotsOff ENTER 2nd DRAW 1 :ClrDraw	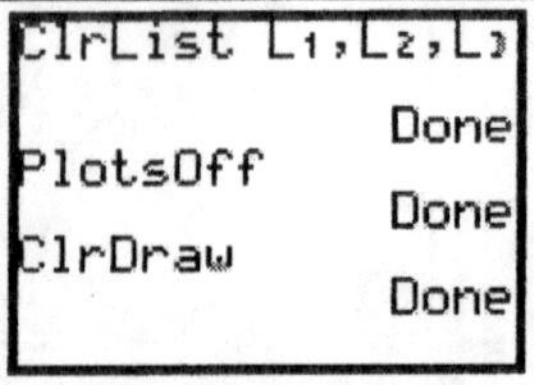 ClrList L1,L2,L3 Done PlotsOff Done ClrDraw Done	Clear all lists. Turn all statistics plots off. Clear all drawings.
STAT 1 :Edit 20 ENTER 60 etc. ▶ 0 ENTER 3 ENTER etc.	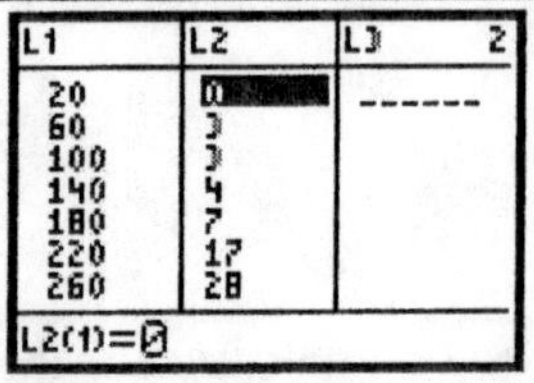 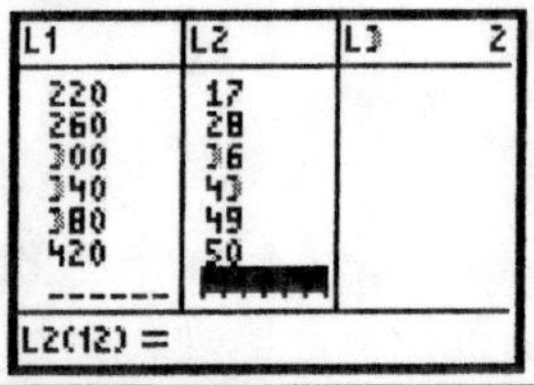 L1: 20, 60, 100, 140, 180, 220, 260 L2: 0, 3, 3, 4, 7, 17, 28 L2(1)=0 L1: 220, 260, 300, 340, 380, 420 L2: 17, 28, 36, 43, 49, 50 L2(12) =	Enter the lower class boundaries and the upper class boundary for the greatest class into L1. Enter the cumulative frequencies into L2.
WINDOW 20 ENTER etc.	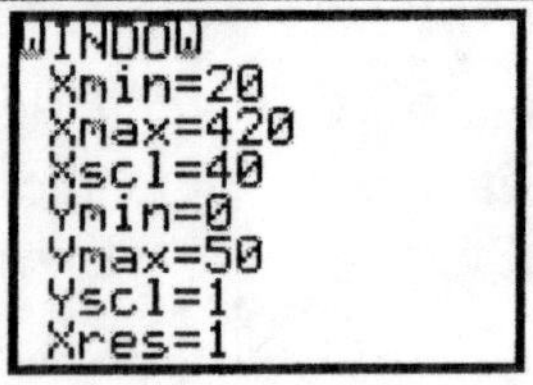 WINDOW Xmin=20 Xmax=420 Xscl=40 Ymin=0 Ymax=50 Yscl=1 Xres=1	Enter the window dimensions: Xmin=lowest class boundary Xmax=highest class boundary Xscl=class width Ymin=0 Ymax=number of pieces of data Yscl=1
2nd STAT PLOT ENTER etc.	 Plot1 Plot2 Plot3 On Off Type: Xlist:L1 Ylist:L2 Mark:	Turn Plot1 on. Set to the line graph with L1 and L2.
GRAPH	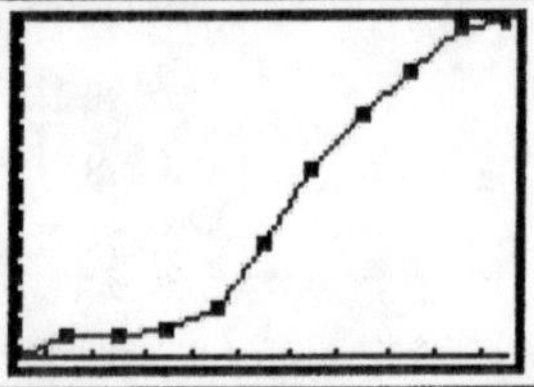	Graph the ogive.
TRACE ▶ ... ▶	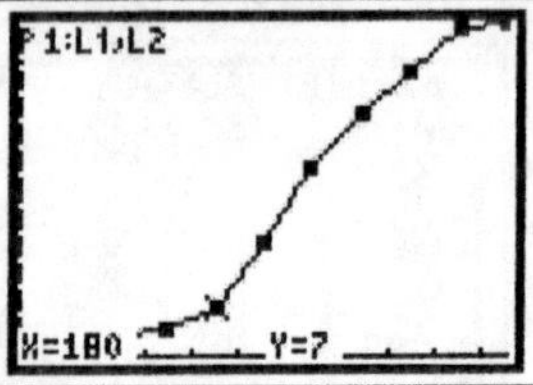 P1:L1,L2 X=180 Y=7	Use TRACE to see the coordinates of the points on the graph.

Example 4

Use the following data to create a time series graph.

Year	1987	1988	1989	1990	1991	1992	1993
Amount	$105	$160	$110	$140	$130	$145	$200

Solution:

Keystrokes	*Screen Display*	*Comments*
[Y=] [CLEAR] [▼] [CLEAR] etc.		Clear all functions from the calculator.
[2nd] [DRAW] [1] :ClrDraw [ENTER]		Clear all drawings from the calculator
[2nd] [STAT PLOT] [4] :PlotsOff [ENTER]		Turn all statistical plots off.
[STAT] [4] :ClrList [2nd] [L1] [,] [2nd] [L2] [ENTER]	ClrDraw Done PlotsOff Done ClrList L1,L2 Done	Clear the lists to be used.
[WINDOW] [1987] [ENTER] [1993] [ENTER] [1] [ENTER] [100] [ENTER] [200] [ENTER] [5] [ENTER]	WINDOW Xmin=1987 Xmax=1993 Xscl=1 Ymin=100 Ymax=200 Yscl=5 Xres=1	Set the graph screen dimensions.
[STAT] [1] :Edit [1987] [ENTER] [1988] [ENTER] [1989] [ENTER] [1990] [ENTER] [1991] [ENTER] [1992] [ENTER] [1993] [ENTER] [►] [105] [ENTER] [160] [ENTER] [110] [ENTER] [140] [ENTER]	L1 L2 L3 2 1988 160 1989 110 1990 140 1991 130 1992 145 1993 200 ------ L2(8) =	Enter the data as bivariate data using the year as the independent variable (input variable) in L1 and the amount as the dependent variable (output variable) in L2.

Keystrokes	Screen	Description
130 ENTER 145 ENTER 200 ENTER		
2nd STAT PLOT 1 :Plot1...Off ENTER		Turn Plot1 on.
▼ ▶ ENTER 2nd L1 ENTER ▼ L2 ENTER ▼ ENTER	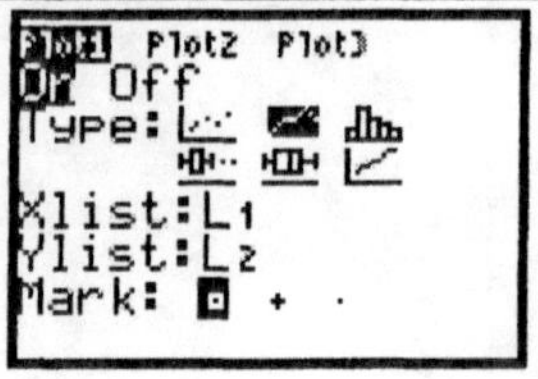	Identify the type of graph, the lists and the mark to use.
GRAPH	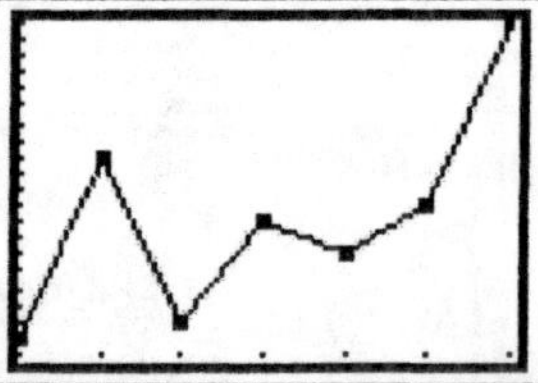	Get the line graph.
TRACE ▶ ... ▶	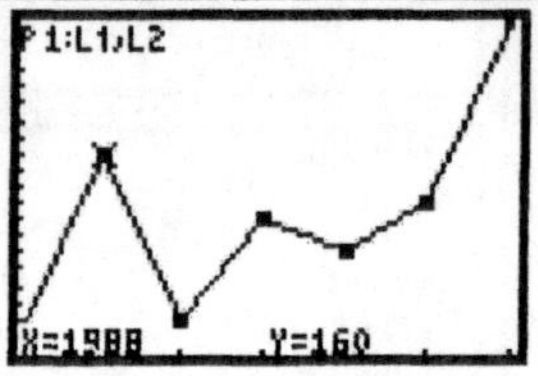	Use TRACE to see the coordinates of the points plotted on the graph.

Example 5

Create a bar chart for the following data:

Profession	Teacher	Doctor	Fireman	Secretary	Nurse	Other
Frequency	20	50	10	30	45	60

Solution:

A bar chart can be drawn using the graphing calculator by assigning a number to represent each category. Assign a number to each profession.

Profession	1	3	5	7	9	11
Frequency	20	50	10	30	45	60

Using odd numbers to identify the categories with an horizontal axis scale of 1 will give spaces between the bars. The data on the horizontal axis will be graphed as 1 ≤ x < 2, 3 ≤ x < 4, etc.

Keystrokes	*Screen Display*	*Comments*
[Y=] [CLEAR] [▼] [CLEAR] etc.		Clear all functions from the calculator.
[2nd] [DRAW] [1] :ClrDraw [ENTER]		Clear all drawings from the calculator
[2nd] [STAT PLOT] [4] :PlotsOff [ENTER]		Turn all statistical plots off.
[STAT] [4] :ClrList [2nd] [L1] [,] [2nd] [L2] [ENTER]	ClrDraw Done PlotsOff Done ClrList L1,L2 Done	Clear the lists to be used.
[WINDOW] [0] [ENTER] [13] [ENTER] [1] [ENTER] [0] [ENTER] [65] [ENTER] [5] [ENTER]	WINDOW Xmin=0 Xmax=13 Xscl=1 Ymin=0 Ymax=65 Yscl=5 Xres=1	Set the graph screen dimensions so there is a space on the left of the leftmost bar and on the right of the rightmost bar.
[STAT] [1] :Edit [1] [ENTER] [3] [ENTER] [5] [ENTER] [7] [ENTER] [9] [ENTER] [11] [ENTER] [►] [20] [ENTER] [50] [ENTER] [10] [ENTER] [30] [ENTER] [45] [ENTER] [60] [ENTER]	L1: 1, 3, 5, 7, 9, 11 L2: 20, 50, 10, 30, 45, 60 L3: L2(1)=20	Enter the data in L1 and L2.

2nd STAT PLOT 1 :Plot1...Off ENTER	STAT PLOTS 1:Plot1...Off L1 L2 2:Plot2...Off L2 L3 3:Plot3...Off L1 L2 4↓PlotsOff	Turn Plot1 on.
▼ ► ► ENTER ▼ 2nd L1 ENTER ▼ 2nd L2 ENTER ▼ ENTER	Plot1 Plot2 Plot3 On Off Type: Xlist:L1 Freq:L2	Identify the type of graph and the lists to use.
GRAPH		Graph.

Example 6

Create a pie graph for the following data:

Profession	Teacher	Doctor	Fireman	Secretary	Nurse	Other
Frequency	20	50	10	30	45	60

Solution:

We will use the list feature of the calculator to calculate the percentages.

Keystrokes	*Screen Display*	*Comments*
STAT 4 :ClrList 2nd L1 , 2nd L2 ENTER	ClrList L1,L2 Done	Clear the lists to be used.

STAT 1 :Edit 20 ENTER 50 ENTER 10 ENTER 30 ENTER 45 ENTER 60 ENTER	L1: 20 50 10 30 45 60 L1(7)=	Enter the data in L1.
2nd QUIT		Return to the Home Screen.
2nd LIST ► ► 5 :sum(2nd L1) ENTER	NAMES OPS MATH 1:min(2:max(3:mean(4:median(5:sum(6:prod(7↓stdDev(sum(L1) 215	Find the sum of the numbers in L1.
2nd L1 ÷ 215 STO► 2nd L2 ENTER STAT 1 :Edit	sum(L1) 215 L1/215→L2 {.0930232558 .2... L1: 20 50 10 30 45 60 L2: .09302 .23256 .04651 .13953 .2093 .27907 L1(1)=20	Divide the numbers in L1 by the total 215 and store in L2. The entire list can be viewed by using the right arrow key or by using the edit feature on the STAT menu.
2nd QUIT 2nd L2 x 360 STO► 2nd L3 ENTER STAT 1 :Edit	L2*360→L3 {33.48837209 83... L1: 20 50 10 30 45 60 L2: .09302 .23256 .04651 .13953 .2093 .27907 L3: 33.488 83.721 16.744 50.233 75.349 100.47 L1(1)=20	Get the Home Screen. Now find the approximate degrees of the circle by multiplying L2 by 360 and storing into L3.

The pie graph must now be drawn by hand using a protractor or estimating where the radii of the circle should be drawn.

EXERCISE SET 2

1. A list of prices of the cheapest pair of jeans at various stores is given below. Construct a histogram having classes of "at least $10 and under $20", "at least $20 and under $30", ... for the data

$32.50	35.00	37.75	42.10	19.00	21.25
11.50	36.20	45.60	39.80	25.55	32.50
33.25	22.59	35.10	30.25	48.00	24.60
20.20	43.80	26.50	35.00	39.20	36.60
38.50	21.10	22.00	23.95	21.00	19.50

2. The ages of a particular state's senators are listed below. Construct a histogram having classes of "at least 20 and under 25", "at least 25 and under 30", "at least 30 and under 35", ... for the data.

52	49	56	69	68	74	41	39	86	20
42	57	56	88	87	37	55	51	38	29
48	25	27	50	53	52	21	22	45	39
80	65	70	55	83	21	35	83	90	41

3. Graph the frequency polygon for the frequency distribution:

Miles Traveled	Number of Cars
at least 3,000 and under 3,500	2
at least 3,500 and under 4,000	3
at least 4,000 and under 4,500	7
at least 4,500 and under 5,000	10
at least 5,000 and under 5,500	15
at least 5,500 and under 6,000	8
at least 6,000 and under 6,500	18
at least 6,500 and under 7,000	2

4. Graph the frequency polygon for the frequency distribution:

Calories consumed per day	Number of Women
at least 1,000 and under 1,200	3
at least 1,200 and under 1,400	6
at least 1,400 and under 1,600	15
at least 1,600 and under 1,800	22
at least 1,800 and under 2,000	25
at least 2,000 and under 2,200	10
at least 2,200 and under 2,400	13
at least 2,400 and under 2,600	8
at least 2,600 and under 2,800	6

5. Construct an ogive for the data in Problem #3.

6. Construct an ogive for the data in Problem #4.

7. Use the data in Problem #3. Change the number of cars in the "6,000 and under 6,500" class from 18 to 10. What effect does this have on the ogive? Explain.

8. Use the data in Problem #4. Change the number of calories in the "2,600 and under 2,800" class from 6 to 18. What effect does this have on the ogive? Explain.

9. Productivity levels in manufacturing a certain product are given below. Construct a line chart for this data.

Month	Jan	Feb	Mar	Apr	May	June	July
Enrollment (number of students)	60	75	70	55	60	76	100

10. Enrollment figures for a certain college are given below. Construct a line chart for this data.

Year	1981	1982	1983	1984	1985	1986	1987
Enrollment (number of students)	400	705	750	650	860	670	900

11. Construct a bar graph for the data in Problem #9 of this exercise set.

12. Construct a bar graph for the data in Problem #10 of this exercise set.

13. Use the data in Problem #9 of this exercise set to:

    (A) Construct a bar graph using percents on the vertical axis. For example, 60 would be replaced by $\frac{60}{60+75+70+55+60+76+100} \times 100$, rounded to the nearest integer.

    (B) Remove July from the data set and repeat part (A). Explain what influence the removal of the July enrollment had on the percents for Jan - June.

14. Use the data in Problem #10 of this exercise set to:

    (A) Construct a bar graph using percents on the vertical axis.

    (B) Remove 1981 from the data set and repeat part (A). Explain what influence the removal of the 1981 enrollment had on the percents for 1982-1987.

NOTES

CHAPTER 3

DATA DESCRIPTION

Example 1

Given the following data:

(A) Compute the arithmetic mean.
(B) Compute the standard deviation.
(C) Find the first quartile.
(D) Find the median.
(E) Find the third quartile.
(F) Find the midrange.
(G) Find the variance.
(H) Find the coefficient of variation.
(I) Construct a boxplot.

Use one decimal place accuracy.

42	21	43	30	38	45	56	34	44	41	47
55	65	36	45	52	49	45	55	39	29	35

Solution (A)-(E):

Keystrokes	*Screen Display*	*Comments*
[CLEAR]		Turn the calculator on and clear the home screen.
[STAT] [4] :ClrList [2nd] [L1] [ENTER]	EDIT CALC TESTS 1:Edit… 2:SortA(3:SortD(4:ClrList 5:SetUpEditor ClrList L1 Done	Clear L1.

	EDIT CALC TESTS 1:Edit... 2:SortA(3:SortD(4:ClrList 5:SetUpEditor	Geet the STAT menu and select [1]:Edit.
[STAT] [1] :Edit... [42] [ENTER] [21] [ENTER] [43] [ENTER] [30] [ENTER] etc.	L1 L2 L3 1 49 45 55 39 29 35 L1(23) =	Enter data. If you make an error, use the arrow keys to highlight the entry. Reenter the number. If you need to delete an entry, use the arrow keys to highlight the entry. Press [DEL] .
[STAT] [▶] :CALC [1] :1-Var Stats	EDIT CALC TESTS 1:1-Var Stats 2:2-Var Stats 3:Med-Med 4:LinReg(ax+b) 5:QuadReg 6:CubicReg 7↓QuartReg	Get the statistics calculation menu from the STAT menu. Select the one-variable statistics option by pressing 1.
[2nd] [L1]	ClrList L_1 Done 1-Var Stats L_1	Specify L1 since we want to get summary numbers for the data in list 1.
[ENTER]	1-Var Stats $\bar{x}$=43 Σx=946 Σx²=42854 Sx=10.17934416 σx=9.945304968 ↓n=22	$\bar{x}$ is the arithmetic mean. The arithmetic mean is 43.0. Sx is the standard deviation. The standard deviation is 10.2.
[▼] [▼] [▼] [▼] [▼]	1-Var Stats ↑n=22 minX=21 Q_1=36 Med=43.5 Q_3=49 maxX=65	Use the down arrow to see the rest of the list. The first quartile is 36.0, the median is 43.5, and the third quartile is 49.0.

Solution (F):

Keystrokes	*Screen Display*	*Comments*
[CLEAR] [(] [VARS] [5] :Statistics... [8] :minX [+] [VARS] [5] :Statistics... [9] :maxX [)] [÷] [2] [ENTER]	VARS Y-VARS 1:Window... 2:Zoom... 3:GDB... 4:Picture... 5:Statistics... 6:Table... 7:String... XY Σ EQ TEST PTS 5↑y 6:Sy 7:σy 8:minX 9:maxX 0:minY A:maxY (minX+maxX)/2 43	The midrange is $\frac{\text{lowest value} + \text{highest value}}{2}$. Get the values from the statistics variables list. Note: The 1-variable statistics needs to have been calculated before finding the midrange. The midrange is 43.0.

Solution (G):

Keystrokes	*Screen Display*	*Comments*
[VARS] [5] :Statistics... [3] :Sx [x^2] [ENTER]	VARS Y-VARS 1:Window... 2:Zoom... 3:GDB... 4:Picture... 5:Statistics... 6:Table... 7:String... XY Σ EQ TEST PTS 1:n 2:x̄ 3:Sx 4:σx 5:ȳ 6:Sy 7↓σy Sx² 103.6190476	To get the variance, recall Sx from the variable statistics list and square it. Note: The 1-variable statistics needs to have been calculated before finding the midrange. The variance is 103.6.

Solution (H):

Keystrokes	*Screen Display*	*Comments*
VARS 5 :Statistics...	VARS Y-VARS 1:Window... 2:Zoom... 3:GDB... 4:Picture... 5:Statistics... 6:Table... 7:String...	Recall the value of the standard deviation from the variable statistics menu.
3 :Sx ÷	XY Σ EQ TEST PTS 1:n 2:x̄ 3:Sx 4:σx 5:ȳ 6:Sy 7↓σy	Divide it by the mean which is also recalled from the variable statistics menu. Multiply the result by 100 to get the coefficient of variation.
VARS 5 :Statistics...	VARS Y-VARS 1:Window... 2:Zoom... 3:GDB... 4:Picture... 5:Statistics... 6:Table... 7:String...	
2 :x̄	XY Σ EQ TEST PTS 1:n 2:x̄ 3:Sx 4:σx 5:ȳ 6:Sy 7↓σy	
x 100 ENTER	Sx/x̄*100 23.67289341	CVar = 23.7%

Solution (I):

Keystrokes	*Screen Display*	*Comments*
2nd STAT PLOT	STAT PLOTS 1:Plot1...Off L1 L2 2:Plot2...Off L2 L3 3:Plot3...Off L1 L2 4↓PlotsOff	Turn all statistical plots off.
4 :PlotsOff ENTER	PlotsOff Done	
2nd STAT PLOT 1 :Plot1...Off ENTER		Turn Plot1 on.

[▼] [▶] [▶] [▶] [ENTER] [▼] [2nd][L1][ENTER] [▼] [1] [ENTER] [▼] [ENTER]	Plot1 Plot2 Plot3 On Off Type: Xlist:L1 Freq:1 Mark: ▪ + ·	Identify the type of graph and the lists to use. The type that is highlighted will indicate outliers if there are any. The frequencies for each item on this list is 1. If an A flashes for Freq press [ALPHA] to change it back to the black rectangle cursor.
[WINDOW] [20][ENTER][70][ENTER] [5][ENTER][0][ENTER] [3][ENTER][1][ENTER]	WINDOW Xmin=20 Xmax=70 Xscl=5 Ymin=0 Ymax=3 Yscl=1 Xres=1	Set the graph screen dimensions. We see in the last screen from part (E) above that the lowest data value is 21 and the greatest is 65. Hence we set the Xmin=20, Xmax=70, Xscl=5. The class width is Xscl. The Ymin=0, Ymax=3, and Yscl=1 since there can be three boxplots displayed at one time on the calculator because there are three different plots possible.
[GRAPH]		Graph. The plot is at the top of the screen since we are using Plot 1.
[TRACE] [▶] ...[▶]	P1:L1 Q3=49	Use [TRACE] and the arrow keys to move the cursor on the screen The numbers displayed will be Xmin, Q_1, Med, Q_3 and Xmax.

Example 2

Compute the mode for the data in Example 1.

42 21 43 30 38 45 56 34 44 41 47
55 65 36 45 52 49 45 55 39 29 35

Solution:

Keystrokes	*Screen Display*	*Comments*
[CLEAR]		Turn the calculator on and clear the home screen.

STAT 4 :ClrList 2nd L1 ENTER	EDIT CALC TESTS 1:Edit… 2:SortA(3:SortD(4:ClrList 5:SetUpEditor ClrList L1 Done	Clear list L1.
STAT 1 :Edit... ENTER 42 ENTER 21 ENTER 43 ENTER 30 ENTER etc.	EDIT CALC TESTS 1:Edit… 2:SortA(3:SortD(4:ClrList 5:SetUpEditor L1 L2 L3 1 49 45 55 39 29 35 L1(23) =	Select the EDIT opeion from the STAT menue in orer to enter the data. Enter data. If you make an error, use the arrow keys to highlight the entry. Reenter the number. If you need to delete an entry, use the arrow keys to highlight the entry. Press DEL . The number will be deleted from the list.
STAT 2 :SortA(2nd L1) ENTER	EDIT CALC TESTS 1:Edit… 2:SortA(3:SortD(4:ClrList 5:SetUpEditor SortA(L1) Done	Select the sort ascending option from the statistics menu. Press 2. Enter the list to be sorted. Close the parentheses and press ENTER .
STAT 1 :Edit ▼ ...▼	L1 L2 L3 1 42 43 44 45 45 45 L1(16) =47	Use the down arrow to examine the list to see if any value occurs more than once. We see that the number 45 occurs three times. This is the mode. No other number occurs three times.

Example 3

Given the following data, compute to one decimal place:

(A) The arithmetic mean.
(B) Standard deviation.
(C) The variance.

Use one decimal place accuracy.

Gallons Sold	Class Midpoints x	Number of Employees (Frequencies) y
80 and less than 90	85	2
90 and less than 100	95	6
100 and less than 110	105	10
110 and less than 120	115	14
120 and less than 130	125	9
130 and less than 140	135	7
140 and less than 150	145	2

Solution (A)-(B):

Keystrokes	*Screen Display*	*Comments*
[CLEAR]		Turn the calculator on and clear the home screen.
[STAT] [4] :ClrList [2nd] [L1] [,] [2nd] [L2] [ENTER]	EDIT CALC TESTS 1:Edit… 2:SortA(3:SortD(4:ClrList 5:SetUpEditor ClrList L1,L2 Done	Clear lists L1 and L2.

STAT 1 :Edit... 85 ENTER 95 ENTER etc. ▶ 2 ENTER 6 ENTER etc.	EDIT CALC TESTS 1:Edit... 2:SortA(3:SortD(4:ClrList 5:SetUpEditor L1 L2 L3 2 95 6 105 10 115 14 125 9 135 7 145 2 L2(8) =	Enter midpoint o each class into L1 and the frequency for each class into L2. The calculator will know which list is the frequency by the command we enter to calculate the statistics.
STAT ▶ :CALC 1 :1-Var Stats	EDIT CALC TESTS 1:1-Var Stats 2:2-Var Stats 3:Med-Med 4:LinReg(ax+b) 5:QuadReg 6:CubicReg 7↓QuartReg	Get the statistics calculation menu. Select the one-variable statistics option.
2nd L1 , 2nd L2	1-Var Stats L1,L2	We want to get summary numbers for the data in L1 with frequencies in L2.
ENTER	1-Var Stats x̄=115.2 Σx=5760 Σx²=674250 Sx=14.77587664 σx=14.6273716 ↓n=50	x̄ is the arithmetic mean. Sx is the standard deviation.

Solution (C):

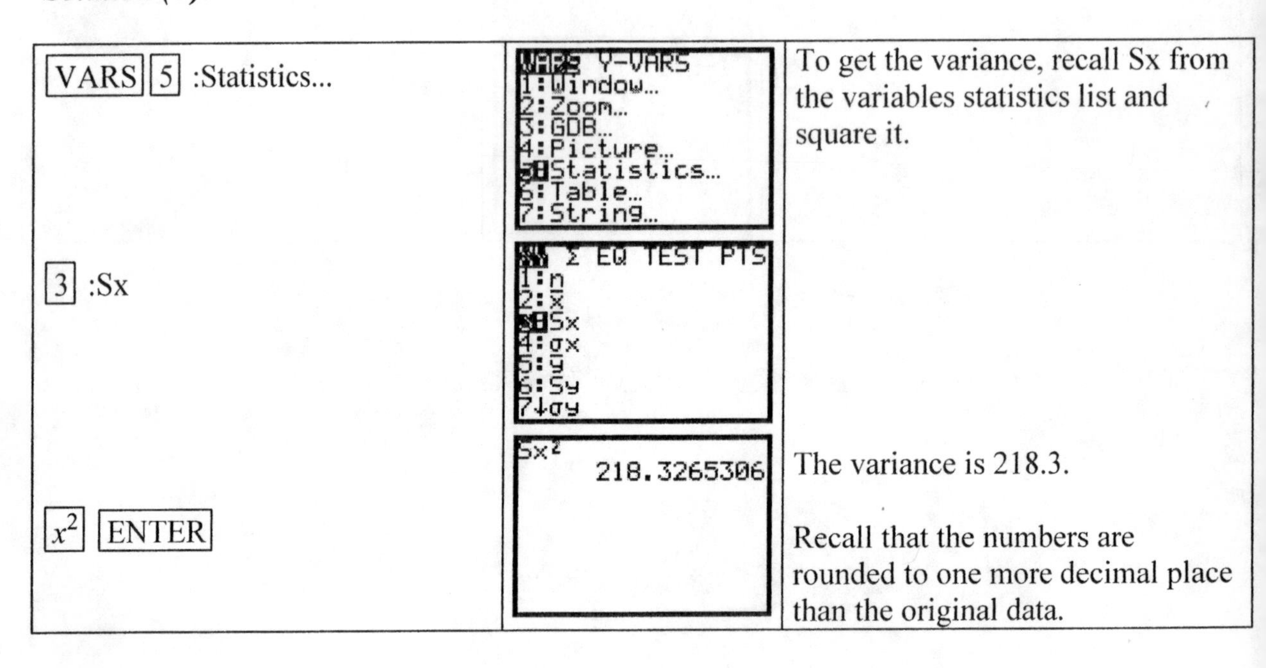

VARS 5 :Statistics... 3 :Sx x^2 ENTER	VARS Y-VARS 1:Window... 2:Zoom... 3:GDB... 4:Picture... 5:Statistics... 6:Table... 7:String... XY Σ EQ TEST PTS 1:n 2:x̄ 3:Sx 4:σx 5:ȳ 6:Sy 7↓σy Sx² 218.3265306	To get the variance, recall Sx from the variables statistics list and square it. The variance is 218.3. Recall that the numbers are rounded to one more decimal place than the original data.

Example 4

Find the grade point average (weighted mean) for the data below.

Course	**Credits (*w*)**	**Grade (*x*)**
Eng Comp I	3	A (4 points)
Intro to Psych	3	C (2 points)
Biology I	4	B (3 points)
Psy Ed	2	D (1 point)

Solution:

Keystrokes	*Screen Display*	*Comments*
CLEAR		Turn the calculator on and clear the home screen.
STAT 4 :ClrList 2nd L1 , 2nd L2 ENTER	EDIT CALC TESTS 1:Edit… 2:SortA(3:SortD(4:ClrList 5:SetUpEditor ClrList L1,L2 Done	Clear lists L1 and L2.
STAT 1 :Edit... (Use the right or left arrow keys to get to L1.) 4 ENTER 2 ENTER etc. ▶ 3 ENTER 3 ENTER etc.	EDIT CALC TESTS 1:Edit… 2:SortA(3:SortD(4:ClrList 5:SetUpEditor L1 L2 L3 2 3 4 3 2 4 3 2 1 L2(5) =	Enter the number of credits (weights) into list L1 and the grade points (data) into list L2.
STAT ▶ :CALC 1 :1-Var Stats	EDIT CALC TESTS 1:1-Var Stats 2:2-Var Stats 3:Med-Med 4:LinReg(ax+b) 5:QuadReg 6:CubicReg 7↓QuartReg	Get the statistics calculation menu. Select the one-variable statistics option.

2nd L2 , 2nd L1	1-Var Stats L2,L 1	We want to get the weighted average for the data in list L2 with weights in list L1.
ENTER	1-Var Stats x̄=2.666666667 Σx=32 Σx²=98 Sx=1.07308674 σx=1.027402334 ↓n=12	x̄ is the weighted mean. The weighted mean is 2.7. The calculator multiplied the numbers in L2 (data) by the numbers in L1 (weights), summed them, and divided the sum by the sum of the numbers in L1 (weights).

Example 5

Create the stem-and-leaf display for the data in Example 1.

42 21 43 30 38 45 56 34 44 41 47
55 65 36 45 52 49 45 55 39 29 35

Solution:

Keystrokes	*Screen Display*	*Comments*
CLEAR		Turn the calculator on and clear the home screen.
STAT 4 :ClrList 2nd L1 ENTER	EDIT CALC TESTS 1:Edit… 2:SortA(3:SortD(4:ClrList 5:SetUpEditor ClrList L1 Done	Clear L1. Or, press the STAT key and press 1: Edit… to get the lists. Press CLEAR and to clear the list. Press 4 :ClrList and 2nd L1 to specify the list to be cleared.
STAT 1 :Edit... ENTER 42 ENTER 21 ENTER 43 ENTER 30 ENTER etc.	EDIT CALC TESTS 1:Edit… 2:SortA(3:SortD(4:ClrList 5:SetUpEditor L1 L2 L3 1 49 45 55 39 29 35 L1(23) =	Get the STAT EDIT L1 again to enter data. If you make an error, use the arrow keys to highlight the entry. Reenter the number. If you need to delete an entry, use the arrow keys to highlight the entry. Press DEL .

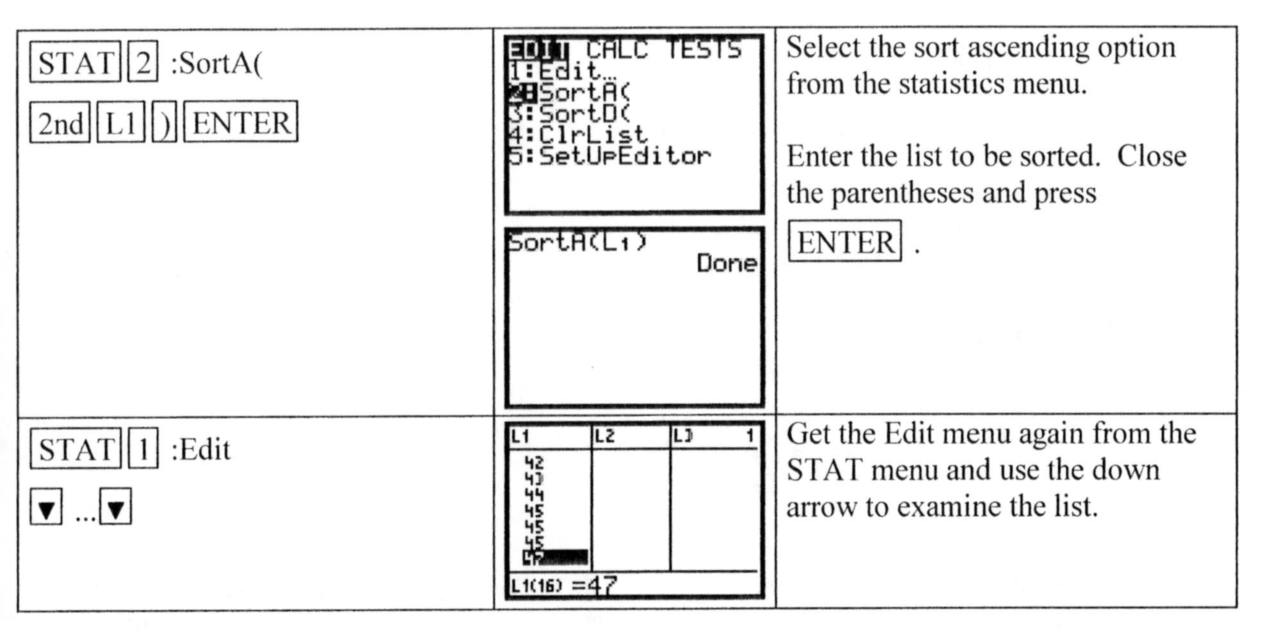

STAT 2 :SortA(2nd L1) ENTER	EDIT CALC TESTS 1:Edit… 2:SortA(3:SortD(4:ClrList 5:SetUpEditor SortA(L1) Done	Select the sort ascending option from the statistics menu. Enter the list to be sorted. Close the parentheses and press ENTER .
STAT 1 :Edit ▼ …▼	L1 L2 L3 1 42 43 44 45 45 45 L1(16) =47	Get the Edit menu again from the STAT menu and use the down arrow to examine the list.

Using the tens digit as the stem and the units digit as the leaf, the stem-and-leaf plot can quickly be written from the sorted list. Since our numbers are from 21 through 65, write 2 through 6 in the stem column. Write the leaf by using the down arrow to exam the sorted list. This must be done by hand with paper and pencil.

Stem	Leaf
2	1 9
3	0 4 5 6 8 9
4	1 2 3 4 5 5 5 7 9
5	2 5 5 6
6	5

The calculator cannot construct a stem-and-leaf plot.

EXERCISE SET 3

1. Given the following data set:

85	76	80	73	72	60	82	89	95
74	83	70	72	88	75	73	74	

(A) Compute the arithmetic mean.
(B) Compute the standard deviation.
(C) Compute the first quartile.
(D) Find the median.
(E) Find the third quartile.
(F) Find the midrange.
(G) Find the variance.
(H) Find the coefficient of variation.
(I) Construct a boxplot.

2. Given the following data set:

5	18	29	24	13	8	22	35	30	50
17	16	20	25	27	21	16	23	25	15

   (A) Compute the arithmetic mean.
   (B) Compute the standard deviation.
   (C) Compute the first quartile.
   (D) Find the median.
   (E) Find the third quartile.
   (F) Find the midrange.
   (G) Find the variance.
   (H) Find the coefficient of variation.
   (I) Construct a boxplot.

3. Use the data in Problem #1. Change the data value 60 to 70. Recalculate the summary measures in parts (A)-(I). What effect did this change have on these summary measures?

4. Use the data in Problem #2. Eliminate the data value 50 from the data set. Recalculate the summary measures in parts (A)-(I). What effect did this change have on these summary measures?

5. Given the following data:

Miles Traveled	Number of Cars
3,000 and under 3,500	2
3,500 and under 4,000	3
4,000 and under 4,500	7
4,500 and under 5,000	10
5,000 and under 5,500	15
5,500 and under 6,000	8
6,000 and under 6,500	18
6,500 and under 7,000	2

   (A) Compute the arithmetic mean.
   (B) Compute the standard deviation
   (C) Find the variance.

6. Given the frequency distribution:

Calories consumed per day	Number of Women
1,000 and under 1,200	3
1,200 and under 1,400	6
1,400 and under 1,600	15
1,600 and under 1,800	22
1,800 and under 2,000	25

2,000 and under 2,200	10
2,200 and under 2,400	7
2,400 and under 2,600	8
2,600 and under 2,800	6

(A) Compute the arithmetic mean.
(B) Compute the standard deviation.
(C) Find the variance.

7. Use the data in Problem #5. Change the number of cars in the "6,000 and under 6,500" class from 18 to 10. Recalculate the summary measures in parts (A)-(C). What effect did this change have on these summary measures?

8. Use the data in Problem #6. Change the number of calories in the "2,600 and under 2,800" class from 6 to 18. Recalculate the summary measures in parts (A)-(C). What effect did this change have on these summary measures?

NOTES

CHAPTER 4

PROBABILITY AND COUNTING RULES

Example 1

Find: (A) 42! (B) ${}_{20}P_{15}$ (C) ${}_{13}C_4$

Solution (A):

Keystrokes	*Screen Display*	*Comments*
CLEAR		Turn the calculator on and clear the home screen.
42 MATH ▶ ▶ ▶ 4 :! ENTER	MATH NUM CPX PRB 1:▶Frac 2:▶Dec 3:³ 4:³√(5:ˣ√ 6:fMin(7↓fMax(MATH NUM CPX PRB 1:rand 2:nPr 3:nCr 4:! 5:randInt(6:randNorm(7:randBin(42! 1.405006118E51	On the math menu, go to the probability menu. The fourth item on the list is the factorial command. The E51 is the calculator notation for scientific notation. E51 means 10^{51} or a 1 with 51 zeros to the right of it. $42! = 1.41x10^{51} = 1.41x1\underbrace{00...000}_{51\ 0's}$

Solution (B):

The quantity can be found by using the definition of permutations $\frac{20!}{(20-15)!}$ or the built-in function for permutations nPr.

Using the definition lf permutations and the factorial command:

Keystrokes	*Screen Display*	*Comments*
CLEAR		Turn the calculator on and clear the home screen.
20 MATH ▶ ▶ ▶ 4 :! ÷ 5 MATH ▶ ▶ ▶ 4 :! ENTER	20!/(20-15)! 2.02741834E16	The factorial symbol is on the math menu on the probability menu. See *Solution (A)* above.

Continuing but using the built-in function for permutations nPr:f

[20] [MATH] [▶] [▶] [▶] [2] :nPr [15] [ENTER]	20!/(20-15)! 2.02741834E16 20 nPr 15 2.02741834E16	Enter the first number.Get the math menu, use the arrow keys to get the probability menu, press 2 to select and get nPr. Enter the second number and press [ENTER].

Solution (C):

The quantity can be found by evaluating the expression $\frac{13!}{4!9!}$ using the factorial command from the PRB menu on the MATH menu or by using the built-in function nCr.

Using the factorial command:

Keystrokes	*Screen Display*	*Comments*
[CLEAR]		Turn the calculator on and clear the home screen.
[13] [MATH] [▶] [▶] [▶] [4]:! [÷] [4] [MATH] [▶] [▶] [▶] [4]:! [ENTER] [÷] [9] [MATH] [▶] [▶] [▶] [4]:! [ENTER]	13!/4! 259459200 Ans/9! 715	The factorial symbol is on the probability menu PRB on the math menu MATH. See *Solution (A)* above. Here we have calculated part of the answer, pressed [ENTER] to get the answer thus far, and then calculated the remainder of the experssion. Notice that the calculator automatically uses the last answer in the next calculation when -, +, ×, ÷ or some other operation.

Using the built-in command:

Keystrokes	*Screen Display*	*Comments*
[13] [MATH] [▶] [▶] [▶] [3] :nCr [4] [ENTER]	13!/4! 259459200 Ans/9! 715 13 nCr 4 715	Enter the first number. Get the math menu, use the arrow keys to get the probability menu, press 3 to select and get nCr. Enter the second number and press [ENTER].

Example 6

Five cards are selected without replacement from a standard poker deck of 52 cards. What is the probability that the first is the ace of clubs, the second is the ace of hearts, the third is the ace of diamonds, the fourth is the ace of spades, and the fifth is a king.

Solution:

There is one way to get the ace of clubs from a deck of 52 cards. So the probability is $\frac{1}{52}$.

There is one way to get the ace of hearts from the remaining 51 cards. So this probability is $\frac{1}{51}$.

There is one way to get the ace of diamonds from the remaining 50 cards. This probability is $\frac{1}{50}$.

There is one way to get the ace of spades from the remaining 49 cards. This probability is $\frac{1}{49}$.

There are four ways to get a king from the remaining 48 cards. This probability is $\frac{4}{48}$.

Keystrokes	*Screen Display*	*Comments*
[CLEAR]		Turn the calculator on and clear the home screen.
[4] [÷] [(] [52] [x] [51] [x] [50] [x] [49] [x] [48] [)] [ENTER]	4/(52*51*50*49*48) 1.282564308E-8	The E-8 is the calculator notation for scientific notation. E-8 means 10^{-8} or a 1 with 7 zeros and a decimal point to the left of it. For example, using the number displayed by the calculator we have: 1.282564308E-8 $= 1.282564308 \times 10^{-8}$ $= 0.00000001282564308$

The answer, in essence, has been rounded to 17 decimal places.

Example 7

Five cards are selected without replacement from a standard poker deck. Find the probability that all of the cards will be diamonds. Round the answer to 8 decimal places.

Solution:

The number of ways to select five cards from thirteen diamonds is ${}_{13}C_5$ since the order of selecting the cards does not matter. The number of ways to select five cards from fifty-two cards is ${}_{52}C_5$. Again, the order of selecting the five cards does not matter. Hence the probability of selecting five diamonds from 52 cards is $\frac{{}_{13}C_5}{{}_{52}C_5}$.

Keystrokes	*Screen Display*	*Comments*
[CLEAR]		Turn the calculator on and clear the home screen.
[13] [MATH] [▶] [▶] [▶] [3]:nCr [5] [÷] [52] [MATH] [▶] [▶] [▶] [3]:nCr [5] [ENTER]	13 nCr 5/52 nCr 5 4.951980792E-4	The combination operation is on the probability menu of the math menu. The answer is 0.00049520.

Example 8

At an art show there are 50 small paintings, 40 middle sized paintings and 20 large paintings. What is the probability that if six paintings are selected at random, three small paintings, two middle sized paintings and one large painting will be selected? Round the answer to four decimal places

Solution:

The number of ways to select three small paintings from 50 is ${}_{50}C_3$.
The number of ways to select two middle sized paintings from 40 is ${}_{40}C_2$.
The number of ways to select one large painting is ${}_{20}C_1 = 20$.

The number of ways to select six paintings from 110 paintings is ${}_{110}C_6$.

Hence the desired probability is: $\frac{({}_{50}C_3)({}_{40}C_2)(20)}{{}_{110}C_6}$.

Keystrokes	*Screen Display*	*Comments*
[CLEAR]		Turn the calculator on and clear the home screen.
[(] [50] [MATH] [▶] [▶] [▶] [3]:nCr [3] [×] [40] [MATH] [▶] [▶] [▶] [20] [)] [÷] [110] [MATH] [▶] [▶] [▶] [3] :nCr [6] [ENTER]	(50 nCr 3*40 nCr 2*20)/110 nCr 6 .1427549859	The combination operation is on the probability menu of the math menu. The probability is 0.1428.

Example 9

An automobile manufacturer has three factories, A, B, and C. They produce 50%, 30%, and 20%, respectively, of a specific model of car. From Factory A for the particular model 30% are white, 40% red, and 30% black. From Factory B 40% are white, 10% red, and 50% black. From Factory C 25% are white, 35% red and 15% black. Find the probabilities for selecting each color of the cars if a car is selected at random from all the cars produced.

Solution:

Let us first organize the numbers into arrays

Factories		
A	B	C
0.50	0.30	0.20

		Color of Car		
		White	Red	Black
Factory	A	0.30	0.40	0.30
	B	0.40	0.10	0.50
	C	0.25	0.35	0.15

This problem can be solved using matrices. Let matrix A = [0.50 0.30 0.20] and let matrix B = . Multiply matrices to find the desired probabilities.

Keystrokes	*Screen Display*	*Comments*
CLEAR		Turn the calculator on and clear the home screen.
MATRX or 2nd MATRX	NAMES MATH EDIT 1:[A] 2:[B] 3:[C] 4:[D] 5:[E] 6:[F] 7↓[G]	On the TI-83 MATRX is on a key cap. On the TI-83+ you need to press 2nd first. Get the matrix menu. Use the right arrows to get the edit menu.
▶ ▶ 1 :[A]	NAMES MATH EDIT 1:[A] 2:[B] 3:[C] 4:[D] 5:[E] 6:[F] 7↓[G]	Get the edit menu for matrices. Select matrix A.
1 ENTER 3 ENTER .50 ENTER .30 ENTER .20 ENTER 2nd QUIT	MATRIX[A] 1 ×3 [.5 .3 .2] 1,3=.2	Enter the dimensions of matrix A as 1 row and 3 columns. Enter the elements of the matrix. Return to the Home Screen.

MATRX or 2nd MATRX ▶ ▶ 1:EDIT 2 :[B]		Get the matrix menu again. Use the right arrows to get the edit menu. Select matrix B.
3 ENTER 3 ENTER .30 ENTER .40 ENTER .30 ENTER .40 ENTER .10 ENTER .50 ENTER .25 ENTER .35 ENTER .15 ENTER 2nd QUIT	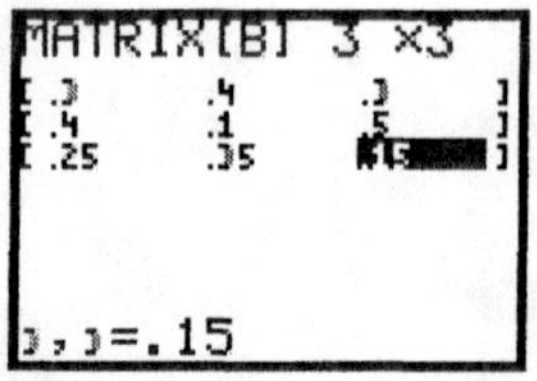 MATRIX[B] 3 ×3 [.3 .4 .3] [.4 .1 .5] [.25 .35 .15] 3,3=.15	Enter the dimensions of the matrix as 3 rows and 3 columns. Enter the elements of the matrix. Return to the Home Screen.
MATRX or 2nd MATRX 1 :[A] x MATRX or 2nd MATRX 2 :[B] ENTER	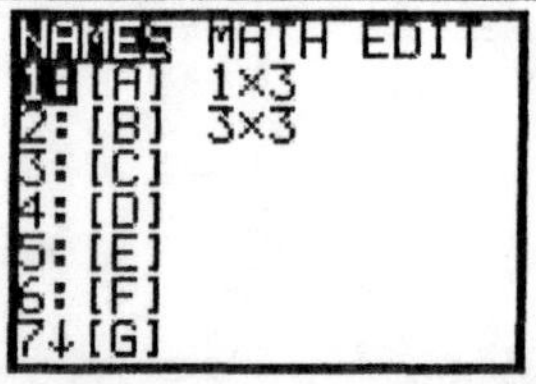 NAMES MATH EDIT 1:[A] 1×3 2:[B] 3×3 3:[C] 4:[D] 5:[E] 6:[F] 7↓[G] 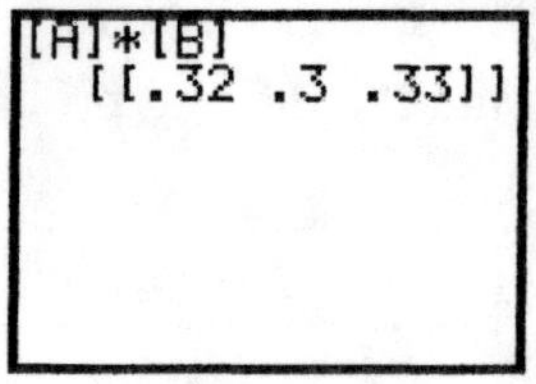 [A]*[B] [[.32 .3 .33]]	Get matrix A from the MATRIX NAMES menu. Multiply by matrix B. The result says that 32% of the cars are white, 30% are red, and 33% are black.

EXERCISE SET 4

1. A restaurant offers five choices of meat, three choices of potatoes, four choices of salad dressing and six choices of beverages. How many different possible meals can be made if a customer must select one item from each category?

2. A person wishes to select an outfit to wear from three choices of hats, four choices of shirts, six choices of pants and ten choices of shoes. How many different possible outfits can be selected if a person must select one item from each category?

3. How many different ways can four people be selected from 20 people on a committee if one must be the chairperson, one the assistant chairperson, one the recording secretary, and one the administrative secretary?

4. A combination lock consisting of three numbers in a sequence can use any of the numbers zero to 29. If no number can be used twice, how many different combinations are possible using three numbers?

5. A coffee specialty shop allows customers to select three different kinds of coffee to be packaged and mailed as a gift item. If there are twenty different varieties available, how many possible selections can be made where any kind can be selected only once?

6. A student can select four courses from a offering of 50 different courses. How many ways can this selection be made? (*Note:* Order of selection is not important.)

7. In a five card hand, what is the probability that four hearts and one spade will be dealt to a person from a standard poker deck? Round answer to four decimal places.

8. What is the probability that a four card hand will have two aces and two kings? Round answer to four decimal places.

9. Six items are selected from a grab bag having 25 red items, 10 blue items and 30 green items. What is the probability that a person will select one red, one blue and 4 green? Round answer to four decimal places.

10. A child has selected 10 books (5 on animals, 3 on games, and 2 on different countries) to check out of a library. However, he is told that only 6 of these books can be checked out. What is the probability that the child will select three animal books, two games books, and 1 on different countries? Round answer to four decimal places.

11. A local restaurant is set up so that customers can select items from a meat counter, a vegetable counter, a salad counter, a dessert counter and a beverage counter. There are 8 meat selections, 5 vegetable selections, 15 salad selections, 5 dessert selections and 10 beverage selections. Of 11 items a customer will select, what is the probability that a customer will select 3 meats, 2 vegetables, 4 salads, 1 dessert and 1 beverage? Round answer to four decimal places.

12. A person buys five raffle tickets. A total of 200 tickets are sold. What is the probability that one of his tickets will be selected in six draws? Round answer to four decimal places.

13. An insurance company classifies drivers as low-risk, medium-risk and high-risk. Of those insured, 60% are low-risk, 30% are medium risk, and 10% are high-risk. After a study, the company finds that during a one-year period, 1% of the low-risk drivers had an accident, and 8% had a driving violation; 5% of the medium-risk drivers had an accident and 10% had a driving violation; and 9% of the high-risk drivers had an accident and 15% had a driving violation. Round answers to four decimal places.

    (A) If a driver is selected at random, what is the probability that the driver had an accident during the year?

    (B) If a driver is selected at random, what is the probability that the driver did not have a a driving violation?

14. A home building company builds 60% three-bedroom homes, 30% four-bedroom homes, and 10% five-bedroom homes. Three features seem to be of importance to buyers: A deck, an attached garage, and a pool. Of the three-bedroom homes, 60% have all three features, 15% have two features, 18% have only one feature, and 7% have no features. Of the four-bedroom homes, 80% have all three features, 10% have two features, 5% have only one feature, and 5% have no features. Of the five-bedroom homes, 95% have all three features, 3% have two features, 2% have one feature, and none have no features. Round answers to four decimal places.

    (A) If a home is selected at random, what is the probability that it has two features?

    (B) If a home is selected at random, what is the probability that it has no features?

CHAPTER 5

DISCRETE PROBABILITY DISTRIBUTIONS

Example 1

One thousand tickets are sold at $1 each for four prizes of $100, $50, $25, and $10. What is the expected value (mean) if a person purchases two tickets?

Gain, *X*	$98	$48	$23	$8	-$2
Probability, *P*(*X*)	$\frac{2}{1000}$	$\frac{2}{1000}$	$\frac{2}{1000}$	$\frac{2}{1000}$	$\frac{992}{1000}$

Solution:

This can easily be found using the list feature in the calculator. Enter the gain into list L1 and the probability into list L2. Find the product of the two lists and sum. The product of the two lists will be placed into list L3.

Keystrokes	*Screen Display*	*Comments*
CLEAR		Turn the calculator on and clear the Home Screen.
STAT 4 :ClrList 2nd L1 , 2nd L2 , 2nd L3 ENTER	EDIT CALC TESTS 1:Edit… 2:SortA(3:SortD(4:ClrList 5:SetUpEditor ClrList L1,L2,L3 Done	Clear the three lists to be used.
STAT 1 :Edit 98 ENTER 48 ENTER 23 ENTER 8 ENTER (-) 2 ENTER ▶ 2 ÷ 1000 ENTER 2 ÷ 1000 ENTER 2 ÷ 1000 ENTER 992 ÷ 1000 ENTER 2nd QUIT	L1 L2 L3 2 98 .002 ------ 48 .002 23 .002 8 .002 -2 .992 ------ L2(6) =	Enter the data into the lists. You can enter 2 ÷ 1000 into the calculator. The calculator will do the division before putting the value in the list. Return to the Home Screen.

2nd L1 x 2nd L2 STO▶ 2nd L3 ENTER	ClrList L1,L2,L3 Done L1*L2→L3 {.196 .096 .046…	Multiply L1 and L2 together and store in L3.
2nd LIST ▶ ▶ 5 :sum(2nd L3) ENTER	NAMES OPS MATH 1:min(2:max(3:mean(4:median(5:sum(6:prod(7↓stdDev(ClrList L1,L2,L3 Done L1*L2→L3 {.196 .096 .046… sum(L3) -1.63	Use the sum(command from the MATH menu on the LIST menu to sum the elements in L3. The sum is -1.63.

Hence, the expected value for a person purchasing two tickets is -$1.63. In the long run the person will lose $1.63 per game.

Example 3

The probability of a defective part in a certain production process is 0.10. A sample of five parts is selected. Find the following:

(A) The probability distribution
(B) Histogram
(C) Mean and standard deviation of the number of defectives in a sample of 5
(D) Probability of exactly 2 defectives.
(E) The probability of two or less defectives.

Express all probabilities rounded to four decimal places.

Solution (A):

The calculator has a built in function to find the probabilities.

Keystrokes	*Screen Display*	*Comments*
CLEAR		Turn the calculator on and clear the Home Screen.
2nd DISTR 0 :binompdf(	DISTR DRAW 6↑χ²pdf(7:χ²cdf(8:Fpdf(9:Fcdf(0:binompdf(A:binomcdf(B↓poissonpdf(	Get the binomial probability density function from the probability distribution menu. Use the arrow keys to select this option or just press 0.

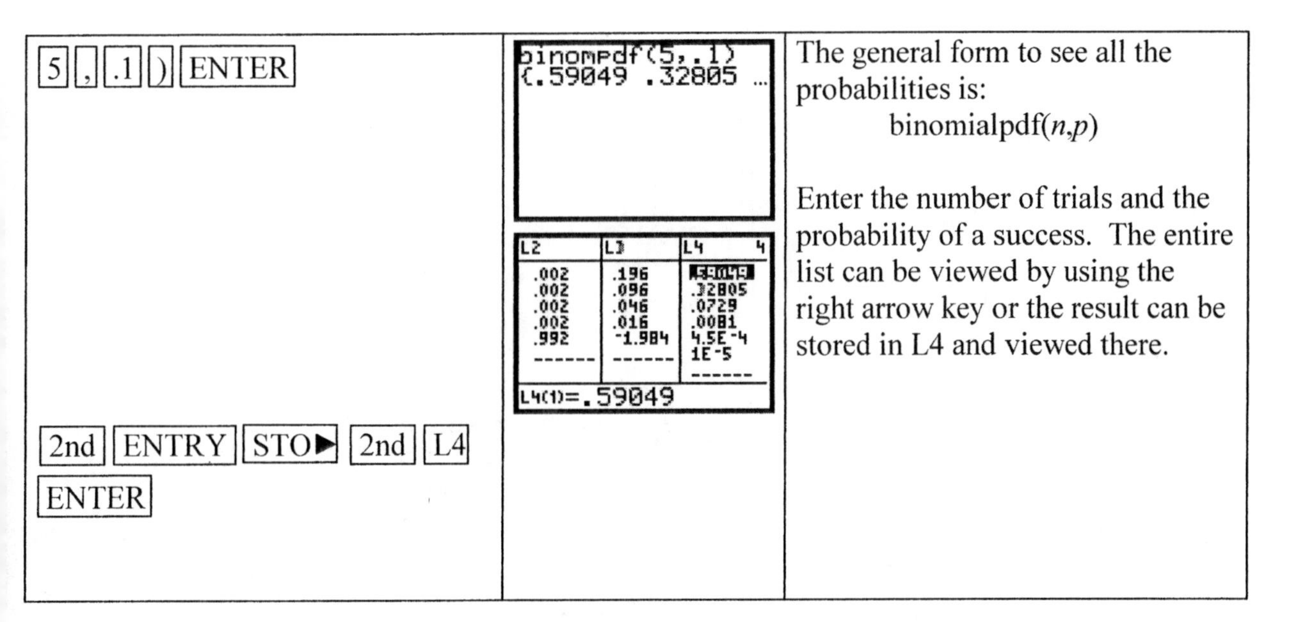

Keystrokes	Screen Display	Comments
[5] [,] [.1] [)] [ENTER] [2nd] [ENTRY] [STO▶] [2nd] [L4] [ENTER]	binompdf(5,.1) {.59049 .32805 ... L2: .002 .002 .002 .002 .992 L3: .196 .096 .046 .016 -1.984 L4: .59049 .32805 .0729 .0081 4.5E-4 1E-5 L4(1)=.59049	The general form to see all the probabilities is: binomialpdf(n,p) Enter the number of trials and the probability of a success. The entire list can be viewed by using the right arrow key or the result can be stored in L4 and viewed there.

To get the probability for a particular X use binomialpdf(n,p,x).

The probabilities could also be calculated using nCr and multiplying by the probabilities as shown in the table below.

X	$P(X)$
0	$_5C_0\,(.1)^0(.9)^5 = 0.5905$
1	$_5C_1\,(.1)^1(.9)^4 = 0.3281$
2	$_5C_2\,(.1)^2(.9)^3 = 0.0729$
3	$_5C_3\,(.1)^3(.9)^2 = 0.0081$
4	$_5C_4\,(.1)^4(.9)^1 = 0.0005$
5	$_5C_5\,(.1)^5(.9)^0 = 0.0000$

Solution (B):

The histogram can be obtained by entering the values of X into list L5 and graphing.

Keystrokes	*Screen Display*	*Comments*
[Y=] [CLEAR] [▼] [CLEAR] ...	Plot1 Plot2 Plot3 \Y1= \Y2= \Y3= \Y4= \Y5= \Y6= \Y7=	Clear all functions from the calculator.

Keystrokes	Screen	Explanation
2nd STAT PLOT 4 :PlotsOff ENTER	STAT PLOTS 1:Plot1…Off L1 1 2:Plot2…Off L1 L2 3:Plot3…Off L1 L2 4↓PlotsOff PlotsOff Done	Turn all statistical plots off.
2nd DRAW 1 :ClrDraw ENTER	DRAW POINTS STO 1:ClrDraw 2:Line(3:Horizontal 4:Vertical 5:Tangent(6:DrawF 7↓Shade(PlotsOff Done ClrDraw Done	Clear all drawing from the calculator
STAT 1 :Edit ► ► ► ► 0 ENTER 1 ENTER etc.	EDIT CALC TESTS 1:Edit… 2:SortA(3:SortD(4:ClrList 5:SetUpEditor	Enter the *x* values into L5.
WINDOW 0 ENTER 6 ENTER 1 ENTER 0 ENTER .6 ENTER .1 ENTER	WINDOW Xmin=0 Xmax=6 Xscl=1 Ymin=0 Ymax=.6 Yscl=.1 Xres=1	Set the graph screen parameters at Xmin=0, Xmax=6, Xscl=1 (class width), Ymin=0, Ymax=0.6, Yscl=0.1, Xres=1. The Xscl is the class width. Xmax=6 so that the bar for *X*=5 will be displayed.
2nd STAT PLOT 1 :Plot1 ENTER	STAT PLOTS 1:Plot1…Off L1 1 2:Plot2…Off L1 L2 3:Plot3…Off L1 L2 4↓PlotsOff	Turn Plot 1 On.
▼ ► ► ENTER ▼ 2nd L5 ENTER 2nd L4 ENTER	Plot1 Plot2 Plot3 On Off Type: Xlist:L5 Freq:L4	Select the Histogram symbol and press enter. Select L5 for the Xlist and L4 for the frequencies.

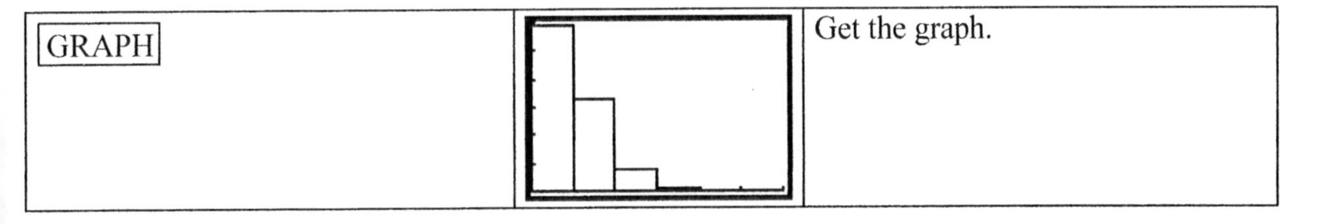

GRAPH		Get the graph.

Solution (C):

Keystrokes	*Screen Display*	*Comments*
CLEAR		Turn the calculator on and clear the Home Screen.
STAT ► ENTER 2nd L5 , 2nd L4 ENTER	1-Var Stats L5,L4 1-Var Stats x̄=.5 Σx=.5 Σx²=.7 Sx= σx=.6708203932 ↓n=1	Select the one-variable statistics option on the statistics calculating menu. The X values are in L5 and the probabilities are in L4. The mean μ is listed as $\bar{x}$. So $\mu = 0.50$ and $\sigma = 0.67$.

Solution (D):

Keystrokes	*Screen Display*	*Comments*
CLEAR		Turn the calculator on and clear the Home Screen.
2nd DISTR 0 :binompdf(5 , .1 , 2) ENTER	binompdf(5,.1,2) .0729	Find an individual probability use: binompdf(n,p,X) The answer is $P(2) = 0.0729$.

Solution (E):

Keystrokes	*Screen Display*	*Comments*
CLEAR		Turn the calculator on and clear the Home Screen.
2nd DISTR ALPHA A	DISTR DRAW 9↑Fcdf(0:binompdf(A:binomcdf(B:poissonpdf(C:poissoncdf(D:geometpdf(E:geometcdf(	Use the down arrow key to select binomial cumulative density function and press ENTER or press ALPHA A :binomcdf(.

[5] [,] [.1] [,] [2] [)] [ENTER]	binomcdf(5,.1,2) .99144	Find cumulative probabilities using: binomcdf(n,p,X) This will find the sum of the probabilities up to and including the probability for x. In this case $P(0)+P(1)+P(2)$. The answer is 0.9914 to four decimal places.

Example 4

Find the probability using the Poisson distribution that

(A) exactly 5 cars arrive at a highway tollgate in a 1-minute period if the average of 3 cars arrive at the tollgate every minute.

(B) at most 5 cars arrive.

Round answers to two decimal places.

Solution (A):

Use the calculator to find the probability $P(5) = \frac{3^5 e^{-3}}{5!}$ by entering:

[3] [^] [5] [×] [2nd] [e^x] [÷] [5] [!]

The factorial function is found on the PRB menu of the MATH menu. See Chapter 4 Example 1 of this manual. The answer is 0.10 or 10%.

Or use the poissonpdf from the distribution menu as shown below.

Keystrokes	*Screen Display*	*Comments*
[CLEAR]		Turn the calculator on and clear the Home Screen.
[2nd] [DISTR] [ALPHA] [B] [3] [,] [5] [)] [ENTER]	DISTR DRAW 9↑Fcdf(0:binompdf(A:binomcdf(B:poissonpdf(C:poissoncdf(D:geometpdf(E:geometcdf(poissonpdf(3,5) .1008188134	Select the poissonpdf(from the DISTR menu. The general form is poissonpdf(λ,X) where λ is the average and X is the number of items. The probability is .10 rounded to two decimal places.

Solution (B):

Keystrokes	*Screen Display*	*Comments*
CLEAR		Turn the calculator on and clear the Home Screen.
2nd DISTR ALPHA C 3 , 5) ENTER	DISTR DRAW 9↑Fcdf(0:binompdf(A:binomcdf(B:poissonpdf(C:poissoncdf(D:geometpdf(E:geometcdf(poissonpdf(3,5) .1008188134 poissoncdf(3,5) .9160820581	Select the poissoncdf(from the DISTR menu. The general form is poissoncdf(λ,X) where λ is the average and X is the number of items. The probability is .92 rounded to two decimal places.

EXERCISE SET 5

1. At Tyler's Tie Shop, Tyler found the probabilities that a customer will buy 0, 1, 2, 3, or 4 ties are as shown below. Find the expected number of ties a person will purchase.

Number of ties, X	0	1	2	3	4
Probability, $P(X)$	0.30	0.50	0.10	0.08	0.02

2. A bank has a drive-through service. The number of customers arriving and the probabilities of arriving during a 15-minute period is distributed as shown below. Find the mean and standard deviation for the distribution.

Number of customers, X	0	1	2	3	4
Probability, $P(X)$	0.12	0.20	0.31	0.25	0.12

3. The probability of a defective part in a sample of six parts is 0.11. Find the following:
   (A) The probability distribution
   (B) Histogram
   (C) Mean and standard deviation of the number of defectives in a sample of 5
   (D) Probability of exactly 2 defectives.
   (E) The probability of two or less defectives.

   Express all probabilities rounded to four decimal places.

4. The probability of a defective part in a sample of eight parts is 0.15. Find the following:
   (A) The probability distribution
   (B) Histogram

(C) Mean and standard deviation of the number of defectives in a sample of 5
(D) Probability of exactly 2 defectives.
(E) The probability of two or less defectives.

Express all probabilities rounded to four decimal places.

5. Given that an average of 4 cars arrive at a tollgate every minute:

(A) Find the probability that exactly 4 cars arrive at a highway tollgate in a 1-minute period.

(B) Find the probability that at least 2 cars arrive at a highway tollgate in a 1-minute period.

Express all probabilities rounded to four decimal places.

6. Given that an average of 4 cars arrive at a tollgate every minute:

(A) Find the probability that exactly 3 cars arrive at a highway tollgate in a 1-minute period.

(B) Find the probability that at least 1 car arrives at a highway tollgate in a 1-minute period.

Express all probabilities rounded to four decimal places.

CHAPTER 6

THE NORMAL DISTRIBUTION

Example 1

Display the normal distribution with $\mu = 35$ and $\sigma = 2$.

Solution:

There are two ways to do this.

METHOD 1

Keystrokes	*Screen Display*	*Comments*
[CLEAR]		Turn the calculator on and clear the Home Screen.
[Y=] [CLEAR] [▼] [CLEAR] …	Plot1 Plot2 Plot3 \Y1= \Y2= \Y3= \Y4= \Y5= \Y6= \Y7=	Clear all functions from the calculator.
[2nd] [STAT PLOT] [4] :PlotsOff [ENTER]	STAT PLOTS 1:Plot1…Off L1 1 2:Plot2…Off L1 L2 3:Plot3…Off L1 L2 4↓PlotsOff PlotsOff Done	Turn all statistical plots off.
[2nd] [DRAW] [1] :ClrDraw [ENTER]	DRAW POINTS STO 1:ClrDraw 2:Line(3:Horizontal 4:Vertical 5:Tangent(6:DrawF 7↓Shade(PlotsOff Done ClrDraw Done	Clear all drawing from the calculator

WINDOW 27 ENTER 43 ENTER 1 (-) .1 ENTER .2 ENTER 1 ENTER 1	WINDOW Xmin=27 Xmax=43 Xscl=1 Ymin=-.1 Ymax=.2 Yscl=1 Xres=1	Set the graph screen dimensions. Add and subtract 4σ from the mean to determine the Xmin and Xmax. The scale should be set so that there will not be more than 20 scale marks. One way to determine this is to calculate $\frac{\text{Xmax-Xmin}}{20}$ and round up. In this case $\frac{43-27}{20} = 0.8$. So we use Xscl=1. Set Ymin -0.1 so that the display will be above the words.
2nd DISTR ► :DRAW 1 :ShadeNorm(27 , 43 , 35 , 2 ENTER	DISTR DRAW 1:ShadeNorm(2:Shade_t(3:Shadeχ²(4:ShadeF(Area=.999937 low=27 up=43	Get the DRAW menu from the DISTR menu. Select ShadeNorm(and enter the numbers needed. The general command is: ShadeNorm(Left,Right,μ,σ) Note: If the standard normal curve is desired, the μ and σ do not have to be entered. ShadeNorm(Left,Right)

METHOD 2

Keystrokes	*Screen Display*	*Comments*
CLEAR		Turn the calculator on and clear the Home Screen.
Y= CLEAR ▼ CLEAR etc.	Plot1 Plot2 Plot3 \Y1= \Y2= \Y3= \Y4= \Y5= \Y6= \Y7=	Clear all functions from the calculator.

2nd STAT PLOT 4 :PlotsOff ENTER	STAT PLOTS 1:Plot1...Off L1 1 2:Plot2...Off L1 L2 3:Plot3...Off L1 L2 4↓PlotsOff PlotsOff Done	Turn all statistical plots off.
2nd DRAW 1 :ClrDraw ENTER	DRAW POINTS STO 1:ClrDraw 2:Line(3:Horizontal 4:Vertical 5:Tangent(6:DrawF 7↓Shade(PlotsOff Done ClrDraw Done	Clear all drawing from the calculator
WINDOW 27 ENTER 43 ENTER 1 (-) .1 ENTER .2 ENTER 1 ENTER 1	WINDOW Xmin=27 Xmax=43 Xscl=1 Ymin=-.1 Ymax=.2 Yscl=1 Xres=1	Set the graph screen dimensions. Add and subtract 4σ from the mean to determine the Xmin and Xmax. The scale should be set so that there will not be more than 20 scale marks. One way to determine this is to calculate $\frac{\text{Xmax-Xmin}}{20}$ and round up. In this case $\frac{43\text{-}27}{20} = 0.8$. So we use Xscl=1. Set Ymin = -0.1 so that the display will be above the words.
Y= 2nd DISTR ENTER X,T,θ,n , 35 , 2) ENTER GRAPH	Plot1 Plot2 Plot3 \Y1=normalpdf(X,35,2) \Y2= \Y3= \Y4= \Y5= \Y6=	Put the normalpdf(into the function list as shown using X as the x value for finding the height of the curve. Get the graph

Example 2

(A) Find the area between $z = 1.18$ and $z = 2.93$ for the standard normal distribution.

(B) Find the probability $P(30.2 < X < 36.8)$ for the normal distribution with $\mu = 35$ and $\sigma = 2$.

(C) If the weight distribution among a population of men is normally distributed with a mean of 180 pounds and a standard deviation of 3.5 pounds, what weight is exceeded by only 6% of the men (the 94th percentile)?

Round all answers to four decimal places.

Solution (A):

Keystrokes	*Screen Display*	*Comments*
[CLEAR]		Turn the calculator on and clear the Home Screen.
[2nd] [DISTR] [2] :normalcdf([1.18] [,] [2.93] [)] [ENTER]	normalcdf(1.18,2.93) .117305288	Get the normal cumulative density function from the DISTR menu and enter the values. $P(30.2 < X < 36.8) = 0.1173$.

Solution (B):

Keystrokes	*Screen Display*	*Comments*
[CLEAR]		Turn the calculator on and clear the Home Screen.
[2nd] [DISTR] [2] :normalcdf([30.2] [,] [36.8] [,] [35] [,] [2] [)] [ENTER]	normalcdf(30.2,36.8,35,2) .8077423796	Get the normal cumulative density function from the DISTR menu and enter the values.

Solution (C):

Keystrokes	*Screen Display*	*Comments*
[CLEAR]		Turn the calculator on and clear the Home Screen.
[2nd] [DISTR] [3] :invNorm([.94] [,] [180] [,] [3.5] [)] [ENTER]	normalcdf(30.2,36.8,35,2) .8077423796 invNorm(.94,180,3.5) 185.4417076	Get the inverse normal cumulative density function from the DISTR menu and enter the values.

Example 3

Display the normal distribution with $\mu = 35$ and $\sigma = 2$ shaded between 30.2 and 36.8.

Solution:

Keystrokes	*Screen Display*	*Comments*
CLEAR		Turn the calculator on and clear the Home Screen.
Y= CLEAR ▼ CLEAR etc.	Plot1 Plot2 Plot3 \Y1= \Y2= \Y3= \Y4= \Y5= \Y6= \Y7=	Clear all functions from the Y= list on the calculator.
2nd STAT PLOT 4 :PlotsOff ENTER	STAT PLOTS 1:Plot1...Off L1 1 2:Plot2...Off L1 L2 3:Plot3...Off L1 L2 4↓PlotsOff PlotsOff Done	Turn all statistical plots off.
2nd DRAW 1 :ClrDraw ENTER	DRAW POINTS STO 1:ClrDraw 2:Line(3:Horizontal 4:Vertical 5:Tangent(6:DrawF 7↓Shade(PlotsOff Done ClrDraw Done	Clear all drawing from the calculator.
WINDOW 27 ENTER 43 ENTER 1 (-) .1 ENTER .2 ENTER 1 ENTER 1	WINDOW Xmin=27 Xmax=43 Xscl=1 Ymin=-.1 Ymax=.2 Yscl=1 Xres=1	Set the graph screen dimensions. Add and subtract 4σ from the mean to determine the Xmin and Xmax. The scale should be set so that there will not be more than 20 scale marks.

> One way to determine this is to calculate $\frac{\text{Xmax-Xmin}}{20}$ and round up. In this case $\frac{43\text{-}27}{20} = 0.8$. So we use Xscl=1. Set Ymin= -0.1 so that the display will be above the words. Use Ymax=0.2 since the height of the normal curve is not greater than this.

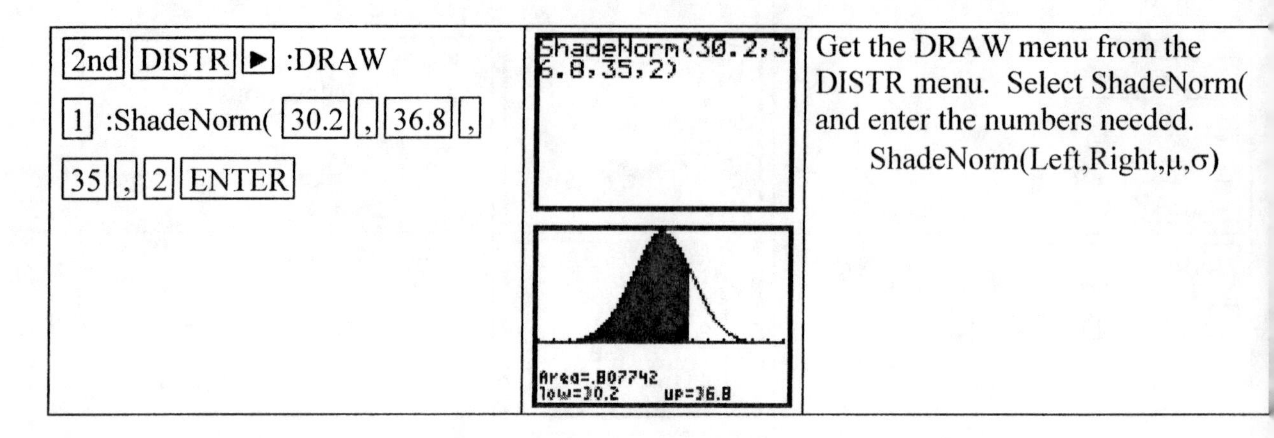

[2nd] [DISTR] [▶] :DRAW [1] :ShadeNorm([30.2] [,] [36.8] [,] [35] [,] [2] [ENTER]	ShadeNorm(30.2,36.8,35,2) Area=.807742 low=30.2 up=36.8	Get the DRAW menu from the DISTR menu. Select ShadeNorm(and enter the numbers needed. ShadeNorm(Left,Right,μ,σ)

Example 4

The average number of gallons of lemonade consumed by football team members during a game is 20 with a standard deviation of 3 gallons. Assume the variable is normally distributed.

(A) When a single game is played, find the probability of using between 15 and 30 gallons.

(B) Find the probability that the average number of gallons consumed from year to year is between 19.0 and 23.5 when 18 games are played per season.

Solution (A):

There are two ways to solve this problem.

METHOD 1

Using the normalcdf(function.

Keystrokes	*Screen Display*	*Comments*
[CLEAR]		Turn the calculator on and clear the Home Screen.
[2nd] [DISTR] [2] :normalcdf([15] [,] [30] [,] [20] [,] [3] [)] [ENTER]	normalcdf(15,30,20,3) .9517805531	Get the normal cumulative density function from the DISTR menu and enter the values.

		The ShadeNorm(command can be used to get a graph of the normal curve with the area shaded in. First the Y= list, plots, and drawings must be cleared or turned off.
[Y=] [CLEAR] [▼] [CLEAR] etc.	Plot1 Plot2 Plot3 \Y1= \Y2= \Y3= \Y4= \Y5= \Y6= \Y7=	Clear all functions from the calculator.
[2nd] [STAT PLOT] [4] :PlotsOff [ENTER]	STAT PLOTS 1:Plot1...Off L1 1 2:Plot2...Off L1 L2 3:Plot3...Off L1 L2 4↓PlotsOff PlotsOff Done	Turn all statistical plots off.
[2nd] [DRAW] [1] :ClrDraw [ENTER]	DRAW POINTS STO 1:ClrDraw 2:Line(3:Horizontal 4:Vertical 5:Tangent(6:DrawF 7↓Shade(PlotsOff Done ClrDraw Done	Clear all drawing from the calculator
[2nd] [DISTR] [▶] :DRAW [1] :ShadeNorm([15] [,] [30] [,] [20] [,] [3] [)] [ENTER]	ShadeNorm(15,30,20,3) Area=.951781 low=15 up=30	The probability can be found using the ShadeNorm(command from the distribution menu even when the graph screen dimensions are not set. The dimensions for the graph shown at the left are Xmin=0, Xmax=7, Xscl=1, Ymin=0, Ymax=1, Yscl=1, Xres=1. Change the dimensions to see the graph.

METHOD 2

Using ShadeNorm(.

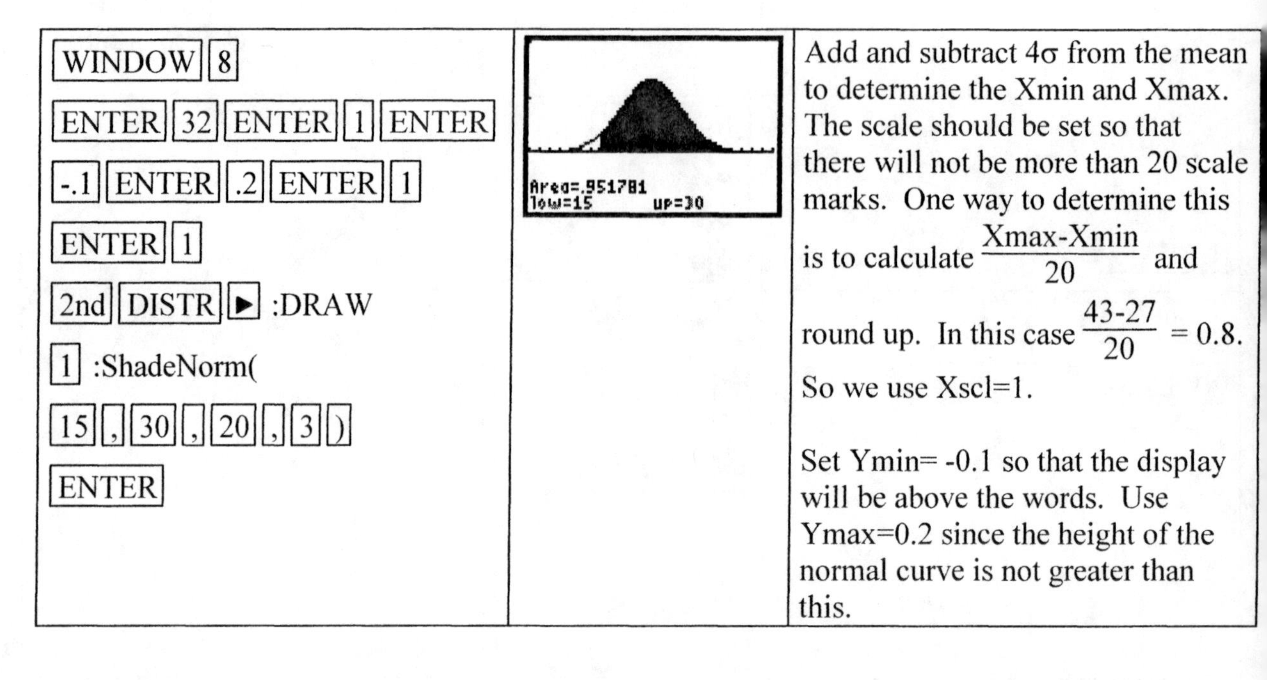

[WINDOW] [8] [ENTER] [32] [ENTER] [1] [ENTER] [-.1] [ENTER] [.2] [ENTER] [1] [ENTER] [1] [2nd] [DISTR] [▶] :DRAW [1] :ShadeNorm([15] [,] [30] [,] [20] [,] [3] [)] [ENTER]	Area=.951781 low=15 up=30	Add and subtract 4σ from the mean to determine the Xmin and Xmax. The scale should be set so that there will not be more than 20 scale marks. One way to determine this is to calculate $\frac{\text{Xmax-Xmin}}{20}$ and round up. In this case $\frac{43-27}{20} = 0.8$. So we use Xscl=1. Set Ymin= -0.1 so that the display will be above the words. Use Ymax=0.2 since the height of the normal curve is not greater than this.

Solution (B):

Find the probability that the average number of gallons consumed from year to year is between 19 and 23 when 18 games are played per season.

Keystrokes	*Screen Display*	*Comments*
[CLEAR]		Turn the calculator on and clear the Home Screen.
[2nd] [DISTR] [2] :normalcdf([15] [,] [30] [,] [20] [,] [3] [)] [ENTER]	normalcdf(15,30, 20,3) .9517805531	Get the normal cumulative density function from the DISTR menu and enter the values.

[WINDOW] [20] [-] [4] [x] [(] [20] [+] [(] [4] [x] [3] [÷] [2nd] [√] [18] [)] [20] [+] [4] [x] [3] [÷] [2nd] [√] [18] [)] [ENTER] [1] [ENTER] [(-)] [.1] [ENTER] [.2] [ENTER] [.1] [ENTER] [1]	WINDOW Xmin=17.171572... Xmax=22.828427... Xscl=1 Ymin=-.1 Ymax=.2 Yscl=.1 Xres=1	Use ShadeNorm(to see how the shape of the graph changes. Set the graph screen dimensions and use ShadeNorm(. Since we are finding the probability of a mean of a sample we must use $\sigma_{\bar{x}} = \frac{3}{\sqrt{18}}$. This can be entered into the calculator as 3÷√ 18). Roundoff error might occur if this quantity is calculated first and entered into the calculator. To set the graph screen dimensions calculate $\mu - 4\sigma_{\bar{x}} = 20 - 4\frac{3}{\sqrt{18}}$ and $\mu + 4\sigma_{\bar{x}} = 20 + 4\frac{3}{\sqrt{18}}$. Enter directly into the calculator or calculate and round off before entering.
[2nd] [DISTR] [▶] :DRAW [1] :ShadeNorm([19] [,] [23.5] [,] [20] [,] [3] [÷] [2nd] [√] [18] [)] [ENTER]	ShadeNorm(19,23.5,20,3/√(18)) Area=.92135 low=19 up=23.5	Notice that the top of the graph is not displayed using Ymax=.2.
[WINDOW] [▼] [▼] [▼] [(-)] [.2] [ENTER] [.7] [ENTER] [.1] [ENTER] [2nd] [QUIT] [2nd] [ENTRY] [ENTER]	WINDOW Xmin=17.171572... Xmax=22.828427... Xscl=1 Ymin=-.2 Ymax=.7 Yscl=.1 Xres=1 Area=.92135 low=19 up=23.5	Change Ymax to .7 to see the entire graph. Also, change Ymin=-.2 so the words do not cover the graph. [2nd] [ENTRY] will recall the most recent command.

Example 5

Suppose we have a finite population of 5000 with a standard deviation of 7.8.

(A) Calculate the population standard deviation of the sample mean if a sample of 100 is taken. Use four decimal places.

(B) Suppose the finite population had a size of 10000. What is the population standard deviation of the sample mean now?

(C) Suppose the finite population had a size of 20000. What is the population standard deviation of the sample mean now?

(D) What do you conjecture the population standard deviation of the sample mean approaches as the size of the population increases?

Solution (A):

We need to calculate $\sigma_{\bar{x}} = \frac{s}{\sqrt{n}}\sqrt{\frac{N-n}{N-1}} = \frac{7.8}{\sqrt{100}}\sqrt{\frac{5000\text{-}100}{5000\text{-}1}}$. Enter the following into the calculator: 7.8 ÷ √ 100 x √ ((5000-100)÷(5000-1)) . The result is 0.7722.

Solution (B):

Use the replay feature of the calculator and change N to 10000. Recall that the command [2nd][ENTRY] displays the last command. After making the change the result is 0.7761. Note that the standard deviation increases when the population size increases.

Solution (C):

Use the replay feature again to change N to 20000. The result is now 0.7781.

Solution (D):

As N gets larger and larger the quantity $\sqrt{\frac{N-n}{N-1}}$ gets closer and closer to 1. Hence the conjecture is that the population standard deviation of the sample means approaches $\sigma_{\bar{x}} = \frac{\sigma}{\sqrt{n}}$. Substituting in the values yields $\sigma_{\bar{x}} = \frac{7.8}{\sqrt{100}} = 0.78$.

Example 6

If a baseball player's batting average is 0.320, find the probability that the player will get at least 26 hits in 100 times at bat. Round the answer to four decimal places.

Solution:

We need to use the continuity correction on the number of hits for approximating the binomial by the normal distribution. We find $P(X \geq 25.5)$ using the normal approximation to the binomial with $\mu = (100)(0.320) = 32$ and $\sigma = \sqrt{(100)(0.320)(0.680)}$. Since we want at least 26 hits we find 1-$P(X<25.5)$.

Keystrokes	*Screen Display*	*Comments*
[CLEAR]		Turn the calculator on and clear the Home Screen.
[1] [-] [2nd] [DISTR] [2] :normalcdf([0] [,] [25.5] [,] [32] [,] [√] [100] [x] [0.320] [x] [0.68] [)] [)] [ENTER]	1-normalcdf(0,25.5,32,√(100*.32*.68)) .9182544435	Get the normal cumulative density function and enter the values. The probability is 0.9183 rounded to four decimal places.

EXERCISE SET 6

1. Display the normal distribution with $\mu = 182$ and $\sigma = 15$.

2. Display the normal distribution with $\mu = 13.2$ and $\sigma = 1.5$.

3. (A) Find the area between $z = -2.38$ and $z = 1.93$ for the standard normal distribution.

   (B) Find the probability $P(51 < X < 168)$ for the normal distribution with $\mu = 60$ and $\sigma = 12$.

   (C) Find the value of X which is exceeded by 17% for the normal distribution with $\mu = 60$ and $\sigma = 12$.

   Round all answers to four decimal places.

4. (A) Find the area between $z = -1.58$ and $z = -0.93$ for the standard normal distribution.

   (B) Find the probability $P(574 < X < 1295)$ for the normal distribution with $\mu = 612$ and $\sigma = 75$.

   (C) Find the value of X which is exceeded by 75% for the normal distribution with $\mu = 612$ and $\sigma = 75$.

   Round all answers to four decimal places.

5. Display the normal distribution with $\mu = 85$ and $\sigma = 22$ shaded between 61 and 128.

6. Display the normal distribution with $\mu = 25$ and $\sigma = 3$ shaded between 26 and 28.

7. Suppose we have a finite population of 8000 with a standard deviation of 23.7.

   (A) Calculate the population standard deviation of the sample mean if a sample of 100 is taken. Use four decimal places.

   (B) Suppose the finite population had a size of 10000. What is the population standard deviation of the sample mean now?

   (C) Suppose the finite population had a size of 20000. What is the population standard deviation of the sample mean now?

   (D) What do you conjecture the population standard deviation of the sample mean approaches as the size of the population increases?

8. Suppose we have a finite population of 9000 with a standard deviation of 43.2.

   (A) Calculate the population standard deviation of the sample mean if a sample of 100 is taken. Use four decimal places.

   (B) Suppose the finite population had a size of 10000. What is the population standard deviation of the sample mean now?

   (C) Suppose the finite population had a size of 20000. What is the population standard deviation of the sample mean now?

   (D) What do you conjecture the population standard deviation of the sample mean approaches as the size of the population increases?

CHAPTER 7

CONFIDENCE INTERVALS AND SAMPLE SIZE

Example 1

For an infinite population with $\sigma = 0.30$, construct an interval estimate with a confidence level of 95% for the population mean, μ, when a sample of size 41 yields $\bar{x} = 1.50$.

Solution (A):

Keystrokes	*Screen Display*	*Comments*
[CLEAR]		Turn the calculator on and clear the Home Screen.
[STAT] [▶] [▶] :TESTS [7] :Zinterval [▶] [ENTER] [▼] [.3] [ENTER] [1.5] [ENTER] [41] [ENTER] [.95] [ENTER] [ENTER]	EDIT CALC TESTS 1:Z-Test… 2:T-Test… 3:2-SampZTest… 4:2-SampTTest… 5:1-PropZTest… 6:2-PropZTest… 7:ZInterval… ZInterval Inpt:Data Stats σ:.3 x̄:1.5 n:41 C-Level:.95 Calculate ZInterval (1.4082,1.5918) x̄=1.5 n=41	Select the tests menu from the statistics menu. Select the ZInterval option. Select [Stats] because the sample mean is given (we are not using data stored in a list). Enter the data. The 95% confidence interval is $1.41 < \mu < 1.59$. The numbers are rounded to the same number of decimal places as the sample mean (one more decimal place than the raw data).

Example 2

Given an infinite population. Construct an interval estimate with a confidence level of 98% for an infinite population the population mean, μ, when a sample of size 21 yields $\bar{x} = 29.83$ and $s = 4.32$.

Solution:

Keystrokes	*Screen Display*	*Comments*
[CLEAR]		Turn the calculator on and clear the Home Screen.

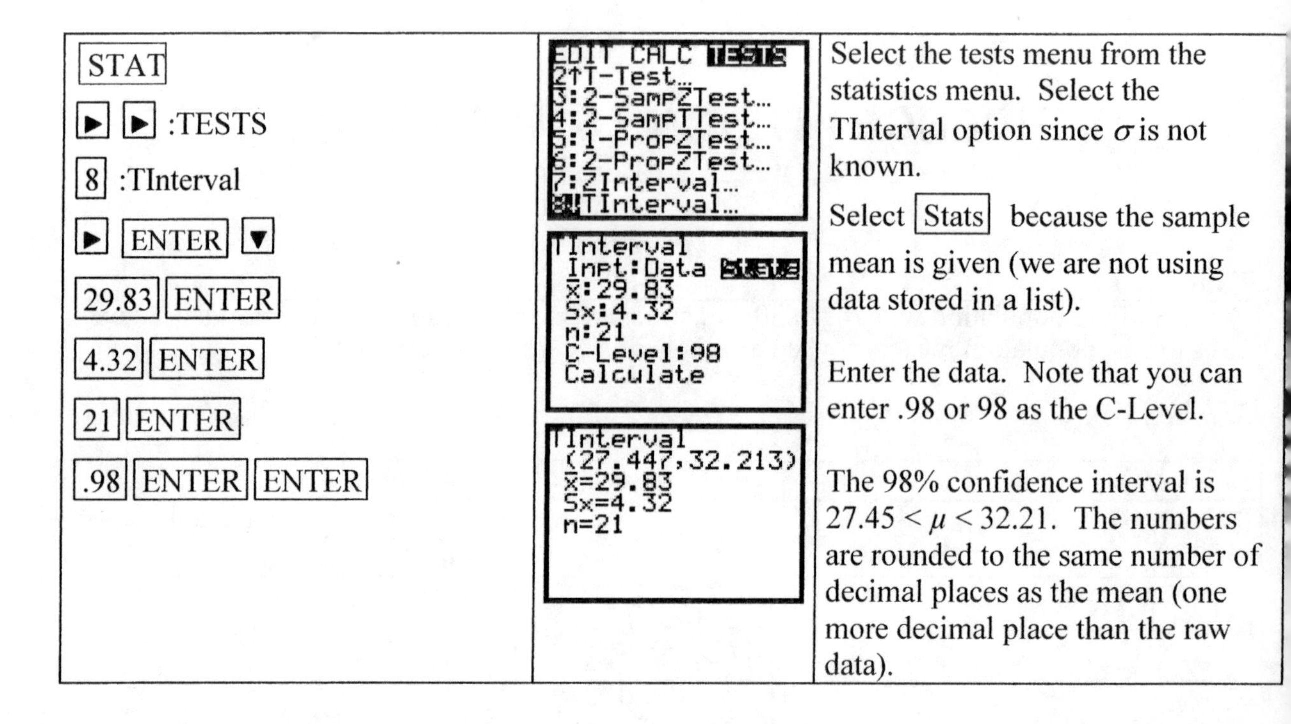

STAT ▶ ▶ :TESTS 8 :TInterval	EDIT CALC TESTS 2↑T-Test... 3:2-SampZTest... 4:2-SampTTest... 5:1-PropZTest... 6:2-PropZTest... 7:ZInterval... 8:TInterval...	Select the tests menu from the statistics menu. Select the TInterval option since σ is not known.
▶ ENTER ▼ 29.83 ENTER 4.32 ENTER 21 ENTER .98 ENTER ENTER	TInterval Inpt:Data Stats x̄:29.83 Sx:4.32 n:21 C-Level:98 Calculate	Select Stats because the sample mean is given (we are not using data stored in a list). Enter the data. Note that you can enter .98 or 98 as the C-Level.
	TInterval (27.447,32.213) x̄=29.83 Sx=4.32 n=21	The 98% confidence interval is $27.45 < \mu < 32.21$. The numbers are rounded to the same number of decimal places as the mean (one more decimal place than the raw data).

Example 3

A sample of 500 nursing applications included 60 from men. Find the 90% confidence interval of the true proportion of men who applied to the nursing program. Use three decimal places in the answer.

Solution:

Keystrokes	*Screen Display*	*Comments*
CLEAR		Turn the calculator on and clear the Home Screen.
STAT ▶ ▶ ALPHA A :1-PropZInt... 60 ▼ 500 ▼	EDIT CALC TESTS 6↑2-PropZTest... 7:ZInterval... 8:TInterval... 9:2-SampZInt... 0:2-SampTInt... A:1-PropZInt... B↓2-PropZInt... 1-PropZInt x:60 n:500 C-Level:90 Calculate 1-PropZInt (.0961,.1439) p̂=.12 n=500	Select the tests menu from the statistics menu. Select the confidence interval option for the proportion for one sample. Enter the data. Note that you do not need the decimal point when entering the confidence level. The point estimate of the population proportion is 0.120. The 90% confidence interval on the population proportion is $0.096 < \mu < 0.144$. The proportions are rounded to three decimal places.

Example 4

The college president asked the statistics teacher to estimate the average age of the students at their college. How large a sample should be taken? The statistics teacher would like to be 99% confident that the estimate will be accurate within one year. From previous study, the standard deviation of the ages is known to be 3 years.

Solution:

We wish to calculate $n = \left(\frac{z_{\alpha/2} \cdot \sigma}{E}\right)^2$. The value of $z_{\alpha/2}$ can be foundusing invNorm(on the distribution menu.

Keystrokes	*Screen Display*	*Comments*
CLEAR		Turn the calculator on and clear the Home Screen.
2nd DISTR 3 :invNorm(.995) ENTER x 3 ÷ 1 ENTER x^2 ENTER	DISTR DRAW 1:normalpdf(2:normalcdf(3:invNorm(4:tpdf(5:tcdf(6:χ²pdf(7↓χ²cdf(invNorm(.995) 2.575829303 invNorm(.995) 2.575829303 Ans*3/1 7.727487909 Ans² 59.71406938	Select the invNorm(option from the distribution men. Enter 0.995 since the area to the left of the number we want from the standard normal distribution is 0.995. The mean of 0 or the standard deviation of 1 is not needed. The calculator will use the standard normal distribution if these values are missing. A sample size of at least 60 students is needed.

EXERCISE SET 7

1. Find a 98% confidence interval on the population mean if a sample of 82 has a mean of 47.8 and the population standard deviation is 3.5.

2. Find a 90% confidence interval on the population mean if a sample of size 41 has a mean of 823.5 and the population standard deviation is 38.2.

3. A study of 36 members of the Central Park Walkers showed that they could walk at an average rate of 2.6 miles per hour. The sample standard deviation is 0.4 miles per hour. Find the 93% confidence interval for the mean of all walkers.

4. Change the number of walkers to 50 in Exercise 8 Problem 1. Find the 93% confidence interval and discuss the effects on the width of the confidence interval when the sample size is increased.

5. In a recent study of 100 people, 85 said that they were dissatisfied with their local elected officials. Find the 98% confidence interval of the true proportion of individuals who are dissatisfied with their local elected officials.

6. In a study of 200 accidents that required treatment in an emergency room, 40% occurred at home. Find the 90% confidence interval of the true proportion of accidents that occur at home.

7. A researcher wishes to estimate, within $25, the true average amount of postage a community college spends each year. If she wishes to be 95% confident, how large a sample is necessary? The standard deviation is known to be $80.

8. Repeat Exercise 8 Problem 7 so that the estimate is within $10.

CHAPTER 8

HYPOTHESIS TESTING

Example 1

In past years the average temperature for a particular two-week period is 67°F. Students at Checkmore School think that the two-week period this year was significantly warmer than typical. They measured the temperature each day at noon for a two-week period.

74 72 69 75 62 64 70 72 78 68 71 62 71 70

Use this data to test the conjecture that the temperature was warmer than 67°F using a significance level of 0.05. Assume the population standard deviation is 5.

Solution:

The hypotheses to be tested are: $H_0 : \mu = 67°F$ and $H_1 : \mu > 67°F$. The data needs to be entered into the calculator and then an hypothesis test completed.

Keystrokes	*Screen Display*	*Comments*
[CLEAR]		Turn the calculator on and clear the Home Screen.
[STAT] [EDIT] [▲] [CLEAR] [▼] [74] [ENTER] [72] [ENTER] etc.	L1 L2 L3 1 78 68 71 62 71 70 L1(15) =	Clear the first list. This is a different way to clear a list than using [STAT] [4] :ClrList [2nd] [L1] . Enter the data into list L1.
[STAT] [►] [►] [1] :Z-Test… [ENTER] [▼] [67] [ENTER] [5] [ENTER] [2nd] [L1] [ENTER] [1] [ENTER] [►] [►] [ENTER] [▼] [ENTER]	EDIT CALC TESTS 1:Z-Test… 2:T-Test… 3:2-SampZTest… 4:2-SampTTest… 5:1-PropZTest… 6:2-PropZTest… 7↓ZInterval… Z-Test Inpt:Data Stats μ0:67 σ:5 List:L1 Freq:1 μ:≠μ0 <μ0 >μ0 Calculate Draw	Select Z-Test from the statistics hypothesis tests menu. Select **Data** since we entered the data into the calculator. Enter the other numbers. Select the greater than option to match H_1

	Z-Test μ>67 z=2.138089935 p=.0162546639 x̄=69.85714286 Sx=4.671635241 n=14	We see that the test value is 2.14 and the probability of observing this value or greater is 0.02 assuming that the null hypothesis is true.
[2nd] [DISTR] [3] :invNorm([.95] [)] [ENTER]	invNorm(.95) 1.644853626	Find the critical value using the inverse normal distribution command. The critical value is 1.64 rounded to two decimal places. Since the test value 2.14 is greater than the critical value it is in the critical region. Hence, the decision is to reject the hypothesis.

There is enough evidence to support the claim that the mean temperature for this two-week period was significantly warmer than typical.

Example 2

The hypotheses for a test on the mean for one sample data are:

$H_0 : \mu = 135$ lbs versus $H_1 : \mu \neq 135$ lbs

The population has a standard deviation of 32 lbs. and is approximately normal. A sample of 38 items has a sample mean of 138 lbs. Should the null hypothesis be rejected at the 0.05 level of significance?

Solution:

Since the sample mean is given there is no raw data to enter into the calculator. The hypotheses can be tested using the one-sample normal test.

Keystrokes	*Screen Display*	*Comments*
[CLEAR]		Turn the calculator on and clear the Home Screen.
[STAT] [▶] [▶] [1] :Z-Test…	EDIT CALC TESTS 1:Z-Test… 2:T-Test… 3:2-SampZTest… 4:2-SampTTest… 5:1-PropZTest… 6:2-PropZTest… 7↓ZInterval…	Select Z-Test from the statistics hypothesis tests menu.

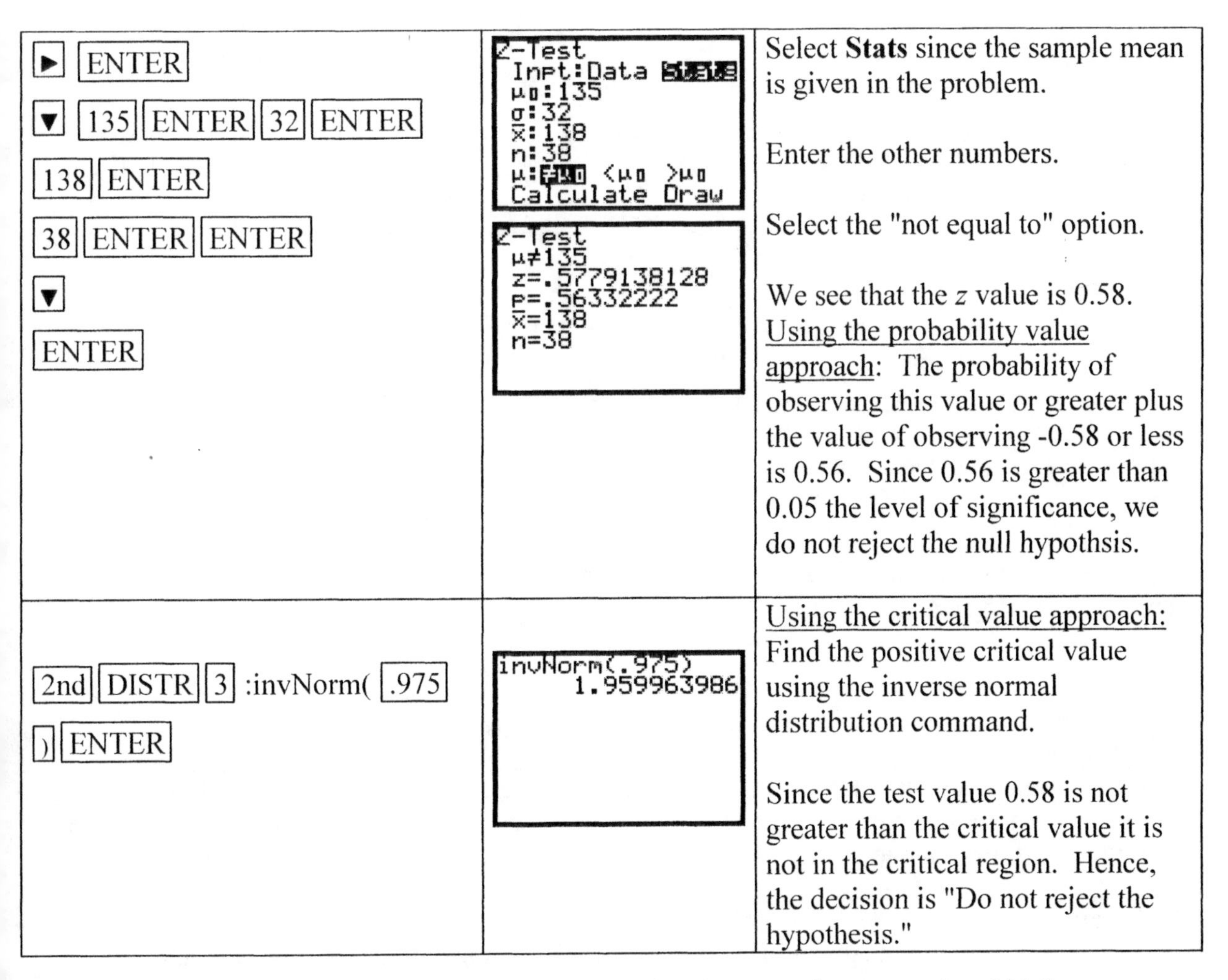

► ENTER ▼ 135 ENTER 32 ENTER 138 ENTER 38 ENTER ENTER ▼ ENTER	Z-Test Inpt:Data Stats μ0:135 σ:32 x̄:138 n:38 μ:≠μ0 <μ0 >μ0 Calculate Draw Z-Test μ≠135 z=.5779138128 p=.56332222 x̄=138 n=38	Select **Stats** since the sample mean is given in the problem. Enter the other numbers. Select the "not equal to" option. We see that the z value is 0.58. Using the probability value approach: The probability of observing this value or greater plus the value of observing -0.58 or less is 0.56. Since 0.56 is greater than 0.05 the level of significance, we do not reject the null hypothsis.
2nd DISTR 3 :invNorm(.975) ENTER	invNorm(.975) 1.959963986	Using the critical value approach: Find the positive critical value using the inverse normal distribution command. Since the test value 0.58 is not greater than the critical value it is not in the critical region. Hence, the decision is "Do not reject the hypothesis."

There is not enough evidence to support the claim that the mean is not equal to 135 lbs.

Example 3

A researcher wishes to test the claim that the average age of lifeguards in Ocean City is greater than 24 years. She selects a sample of 36 lifeguards and finds the mean age of the sample to be 24.7 years, with a standard deviation of 2 years. Is there evidence to support the claim at $\alpha =$ 0.05? Use the P-Value (probability value) method for hypothesis testing.

Solution:

Keystrokes	*Screen Display*	*Comments*
CLEAR		Turn the calculator on and clear the Home Screen.
STAT ► ► 1 :Z-Test...	EDIT CALC TESTS 1:Z-Test... 2:T-Test... 3:2-SampZTest... 4:2-SampTTest... 5:1-PropZTest... 6:2-PropZTest... 7↓ZInterval...	Select the tests menu from the statistics menu. Select the one sample z test.

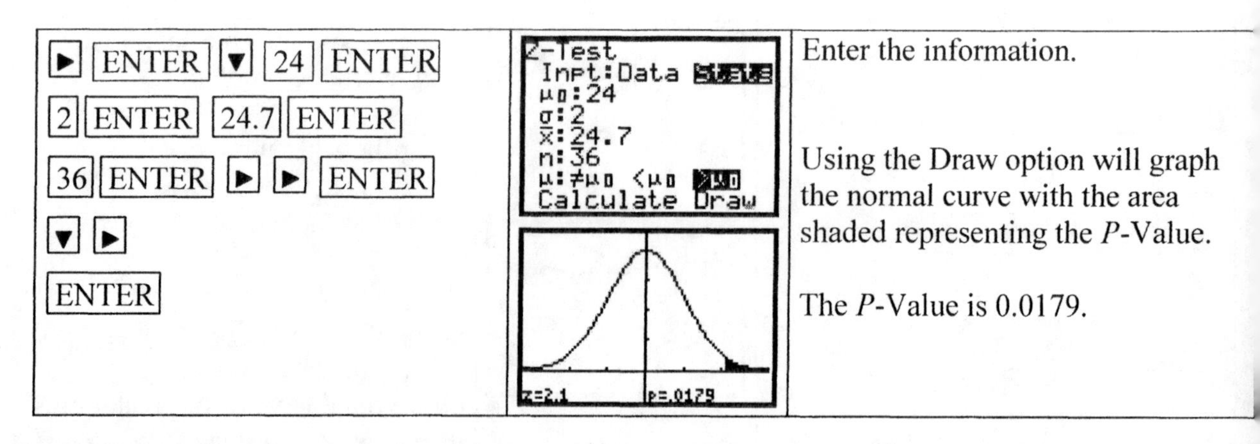

▶ ENTER ▼ 24 ENTER 2 ENTER 24.7 ENTER 36 ENTER ▶ ▶ ENTER ▼ ▶ ENTER	Z-Test Inpt:Data Stats μ0:24 σ:2 x̄:24.7 n:36 μ:≠μ0 <μ0 >μ0 Calculate Draw z=2.1 p=.0179	Enter the information. Using the Draw option will graph the normal curve with the area shaded representing the *P*-Value. The *P*-Value is 0.0179.

Since the *P*-Value is less than the level of significance, the decision is to reject the null hypothesis.

There is enough evidence to support the claim that the average age of lifeguards in Ocean City is greater than 24 years.

Example 4

A job placement director claims that the average starting salary for nurses is \$24,000. A sample of 10 nurses has a mean of \$23,450 and a standard deviation of \$400. Is there enough evidence to reject the director's claim at $\alpha = 0.05$? Use the *P*-Value method for hypothesis testing.

Solution:
The sample information is given. Hence, this information is entered into the calculator and the one-sample *t*-test is used.

Keystrokes	*Screen Display*	*Comments*
CLEAR		Turn the calculator on and clear the Home Screen.
STAT ▶ ▶ 2:T-Test…	EDIT CALC TESTS 1:Z-Test… 2:T-Test… 3:2-SampZTest… 4:2-SampTTest… 5:1-PropZTest… 6:2-PropZTest… 7↓ZInterval…	Select the tests menu from the statistics menu. Select the one-sample *t*-test.
▶ ENTER ▼ 24000 ENTER 23450 ENTER 400 ENTER 10 ENTER ENTER	T-Test Inpt:Data Stats μ0:24000 x̄:23450 Sx:400 n:10 μ:≠μ0 <μ0 >μ0 Calculate Draw	Enter the information.

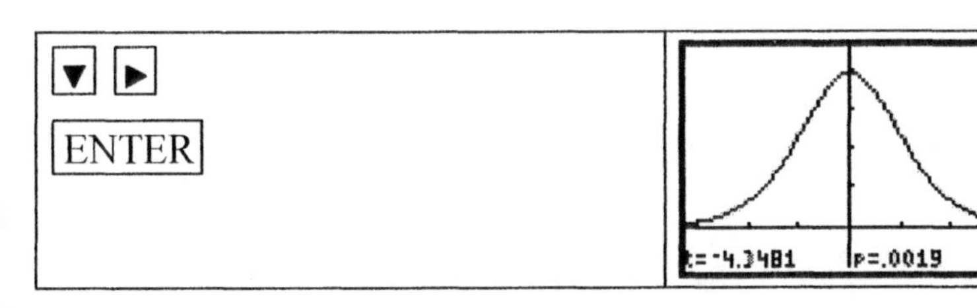

Using the Draw option will graph the t distribution with the area shaded representing the P-Value.

The P-Value is 0.0019.

The decision is to reject the null hypothesis since the P-Value is less than the level of significance.

There is enough evidence to reject the claim that the starting salary of nurses is $24,000.

Example 5

An educator estimates that the dropout rate for seniors at high schools in Ohio is 15%. Last year, 38 seniors from a random sample of 200 Ohio seniors withdrew. At $\alpha = 0.05$, is there enough evidence to reject the educator's claim? Use the P-Value method for hypothesis testing.

Solution:

The sample information is given. Hence, this information is entered into the calculator and the one-sample test for proportions is used.

Keystrokes	*Screen Display*	*Comments*
CLEAR		Turn the calculator on and clear the Home Screen.
STAT ▶ ▶ 5 :1-PropZTest	EDIT CALC TESTS 1:Z-Test... 2:T-Test... 3:2-SampZTest... 4:2-SampTTest... 5:1-PropZTest... 6:2-PropZTest... 7↓ZInterval...	Select the tests menu from the statistics menu. Select the one sample proportion test.
.15 ENTER 38 ENTER 200 ENTER ENTER	1-PropZTest p0:.15 x:38 n:200 prop≠p0 <p0 >p0 Calculate Draw	Enter the information.
▼ ENTER	1-PropZTest prop≠.15 z=1.584236069 p=.1131400131 p̂=.19 n=200	Using the Calculate option yields the P-Value to be 0.113. Note that $\hat{p}$ is also given and is 0.19.

The decision is not to reject the null hypothesis since the P-Value is not less than the level of significance.

There is not enough evidence to reject the claim that the dropout rate for seniors in high schools in Ohio is 15%.

Example 6

(A) Graph the χ^2 distribution with 8 degrees of freedom.

(B) Graph the χ^2 distribution with 15 degrees of freedom with the area between 6.58 and 10.92 shaded.

Solution (A):

Keystrokes	*Screen Display*	*Comments*
CLEAR		Turn the calculator on and clear the Home Screen.
WINDOW 0 ENTER 20 ENTER 1 ENTER (-) .05 ENTER .15 ENTER .1 ENTER	WINDOW Xmin=0 Xmax=20 Xscl=1 Ymin=-.05 Ymax=.15 Yscl=.1 Xres=1	Set the graph screen dimensions. You may need to try different settings to get the display you want.
Y= CLEAR 2nd DISTR 6 :χ^2pdf(X,T,θ,*n* , 8) ENTER	Plot1 Plot2 Plot3 \Y1=χ²pdf(X,8) \Y2= \Y3= \Y4= \Y5= \Y6= \Y7=	Clear any functions in the Y= list. Get the Chi-square distribution and enter 8 degrees of freedom.
GRAPH		Graph the function.

Solution (B):

Keystrokes	*Screen Display*	*Comments*
CLEAR		Turn the calculator on and clear the Home Screen.
Y= CLEAR		Clear any functions in the Y= list.
WINDOW 0 ENTER 30 ENTER 1 ENTER (-) .05 ENTER .15 ENTER .1 ENTER	WINDOW Xmin=0 Xmax=30 Xscl=1 Ymin=-.05 Ymax=.15 Yscl=.1 Xres=1	Set the graph screen dimensions.

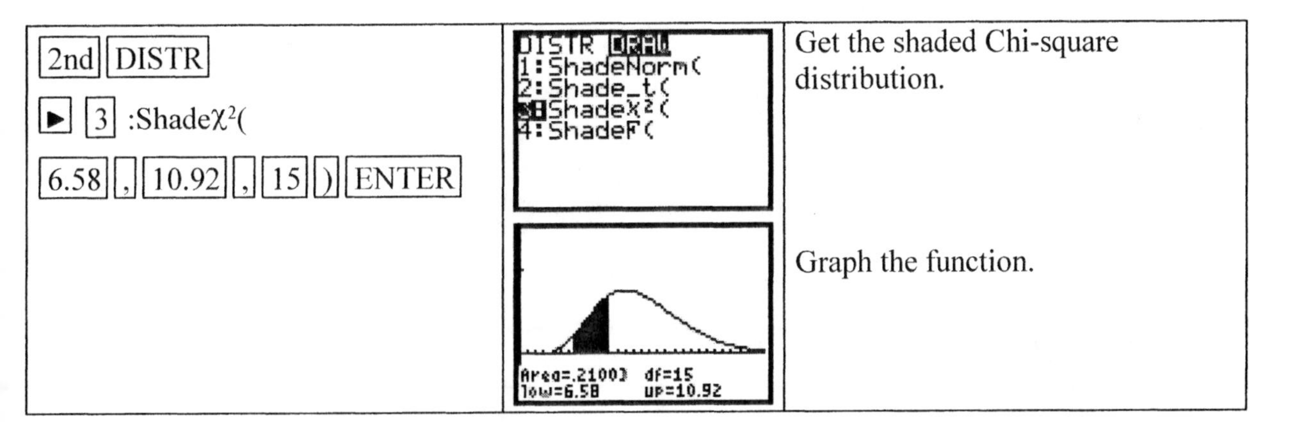

2nd DISTR ▶ 3 :Shadeχ^2(6.58 , 10.92 , 15) ENTER	DISTR DRAW 1:ShadeNorm(2:Shade_t(3:Shadeχ^2(4:ShadeF(Area=.21003 df=15 low=6.58 up=10.92	Get the shaded Chi-square distribution. Graph the function.

Example 7

A researcher knows from past studies that the standard deviation of the time it takes to inspect a car is 16.8 minutes. A sample of 24 cars is selected and inspected. The standard deviation was 12.5 minutes. At $\alpha = 0.05$, is there enough evidence to conclude that the standard deviation of the population has changed? Use the *P*-Value method for hypothesis testing.

Solution:

The *P*-Value can be found using the calculator once the test value is calculated.

Keystrokes	*Screen Display*	*Comments*
CLEAR		Turn the calculator on and clear the Home Screen.
(24 - 1) x 12.5 x^2 ÷ 16.8 x^2 ENTER	(24-1)*12.5²/16.8² 12.73295777	Calculate the test value.
2nd DISTR 7 :χ^2cdf(0 , 12.733 , 23) ENTER	DISTR DRAW 1:normalpdf(2:normalcdf(3:invNorm(4:tpdf(5:tcdf(6:χ^2pdf(7:χ^2cdf((24-1)*12.5²/16.8² 12.73295777 χ^2cdf(0,12.733,23) .0424560752	Find the probability between 0.000 and 12.733. $P(0 < \chi^2 < 12.733) = 0.042$.

Since this is a two-tailed test the probability must be doubled. Hence the *P*-Value is 0.084. The *P*-Value is greater than the level of significance. Hence, the decision is not to reject the null hypothesis.

There is not enough evidence to support the claim that the standard deviation has changed.

EXERCISE SET 8

1. The hypotheses for a test of hypothesis are:

   $H_0: \mu = 23$ lbs. versus $H_1: \mu < 23$ lbs.

   The population has a standard deviation of 3.5 lbs. and is approximately normally distributed. The sample mean is $\bar{x} = 21.0$ lbs.

   (A) Calculate to two decimal places the test value if the sample consisted of 35 items. Use four decimal places.
   (B) Suppose the sample consisted of 70 items with the same standard deviation and sample mean. Find the test value.
   (C) Compare Parts (A) and (B). What happens to the test value as the sample size increases? Does it double in size as the sample size doubles?

2. The hypotheses for a test of hypothesis are:

   $H_0: \mu = 65$ lbs. versus $H_1: \mu \neq 65$ lbs.

   The population has a standard deviation of 5.3 lbs. and is approximately normally distributed. The sample mean is $\bar{x} = 62$.

   (A) Calculate to two decimal places the test value if the sample consisted of 37 items.
   (B) Suppose the sample consisted of 74 items with the same standard deviation and sample mean. Find the test value.
   (C) Compare Parts (A) and (B). What happens to the test value as the sample size increases? Does it double in size since the sample size is doubled?

3. Refer to Problem 1 above.
   (A) Should the null hypothesis be rejected at $\alpha = 0.01$ in Problem 1(A)?
   (B) Should the null hypothesis be rejected $\alpha = 0.01$ in Problem 1(B)?
   (C) Is the decision different in (A) and (B)? Why?

4. Refer to Problem 2 above.
   (A) Should the null hypothesis be rejected at $\alpha = 0.05$ in Problem 2(A)?
   (B) Should the null hypothesis be rejected at $\alpha = 0.05$ in Problem 2(B)?

(C) Is the decision different in (A) and (B)? Why?

5. Refer to Example 1 of this chapter. We notice that there was one very warm day with a temperature of 78°F. Would the null hypothesis be rejected if this value was removed from the data set?

6. Refer to Example 1 of this chapter. Suppose the warm day recorded with temperature 78°F was recorded in error and it should have been 68°F. Would the null hypothesis be rejected now?

7. A high school counselor wishes to test the theory that the average age of the dropouts in her school district is 16.3 years. She samples 12 recent dropouts and finds that their mean age is 16.9 years and the standard deviation is 0.5 years. At $\alpha = 0.05$, is the theory refutable?

8. Students at a local college claim that the textbooks are more expensive this semester than they have ever been. In the past, the average cost per student per semester for textbooks was $178.35. Test the hypothesis that the cost of textbooks has increased using the following data:

$166.50	$155.00	$178.95	$189.34	$133.55	$125.44	$164.53	$189.30	$187.76
$ 98.76	$134.55	$178.98	$168.55	$198.78	$145.34	$155.73	$177.65	$190.43
$155.88	$122.55	$190.55	$187.55	$134.35	$131.45	$145.33	$136.57	$137.89
$133.56	$125.44	$167.89	$158.93	$136.77	$135.44			

9. A dietitian read in a survey that at least 60% of adults eat hamburgers for lunch at least three times during the five-day work week. To test this claim, she selected a random sample of 65 adults and asked them how many days a week they ate hamburgers. 35 people said they did. At $\alpha = 0.01$, do her results refute the survey results?

10. A radio manufacturer claims that 65% of teenagers 13-16 years old have their own portable CD players. A researcher wishes to test the claim and selects a random sample of 80 teenagers. She finds that 57 have their own CD players. At $\alpha = 0.05$ should the claim be rejected?

11. A film editor thinks that the standard deviation for the number of minutes in a video is 3.4 minutes. A sample of 24 videos has a standard deviation of 4.2 minutes. Find the *P*-value of this test.

12. A manufacturer claims that the standard deviation of the drying time of a certain type of paint is 18 minutes. A sample of 8 test panels produced a standard deviation of 20 minutes. Find the *P*-Value of this test.

13. Refer to Problems 1 and 3 above. Find the 99% confidence interval for Problem 3(A) and 3(B). Do the results agree with your decisions in Problems 3(A) and 3(B)? Explain.

14. Refer to Problems 2 and 4 above. Find the 95% confidence interval for Problems 4(A) and 4(B). Do the results agree with your decision in Problems 4(A) and 4(B)? Explain.

15. (A) Graph the χ^2 distribution with 10 degrees of freedom.

(B) Graph the χ^2 distribution with 18 degrees of freedom with the area between 7.52 and 12.03 shaded.

16. (A) Graph the χ^2 distribution with 15 degrees of freedom.

    (B) Graph the χ^2 distribution with 8 degrees of freedom with the area between 3.58 and 8.99 shaded.

NOTES

CHAPTER 9

TESTING THE DIFFERENCE BETWEEN TWO MEANS, TWO VARIANCES, AND TWO PROPORTIONS

Example 1

A survey found that the average hotel room rate in New Orleans is $88.42 and the average room rate in Phoenix is $80.61. Assume that the data were obtained from two samples of 50 hotels in New Orleans and 35 hotels in Phoenix and that the population standard deviations were $5.62 and $5.83, respectively.

Test the hypothesis

$H_0: \mu_N = \mu_P$ versus $H_1: \mu_N \neq \mu_P$

Use the probability value approach.

Solution:

The information is entered into the calculator and a two-sample test on sample means is performed using 2-SampZTest... .

Keystrokes	*Screen Display*	*Comments*
CLEAR		Turn the calculator on and clear the Home Screen.
STAT ► ► 3 :2-SampZTest...	EDIT CALC TESTS 1:Z-Test... 2:T-Test... 3:2-SampZTest... 4:2-SampTTest... 5:1-PropZTest... 6:2-PropZTest... 7↓ZInterval...	Select two-sample *z* test from the statistics hypothesis tests menu.
► ENTER ▼ 5.62 ENTER 5.83 ENTER 88.42 ENTER 50 ENTER 80.61 ENTER 35 ENTER ENTER	2-SampZTest Inpt:Data Stats σ1:5.62 σ2:5.83 x̄1:88.42 n1:50 x̄2:80.61 ↓n2:35 2-SampZTest ↑σ2:5.83 x̄1:88.42 n1:50 x̄2:80.61 n2:35 μ1:≠μ2 <μ2 >μ2 Calculate Draw	Enter the information. Select the not-equal-to option.

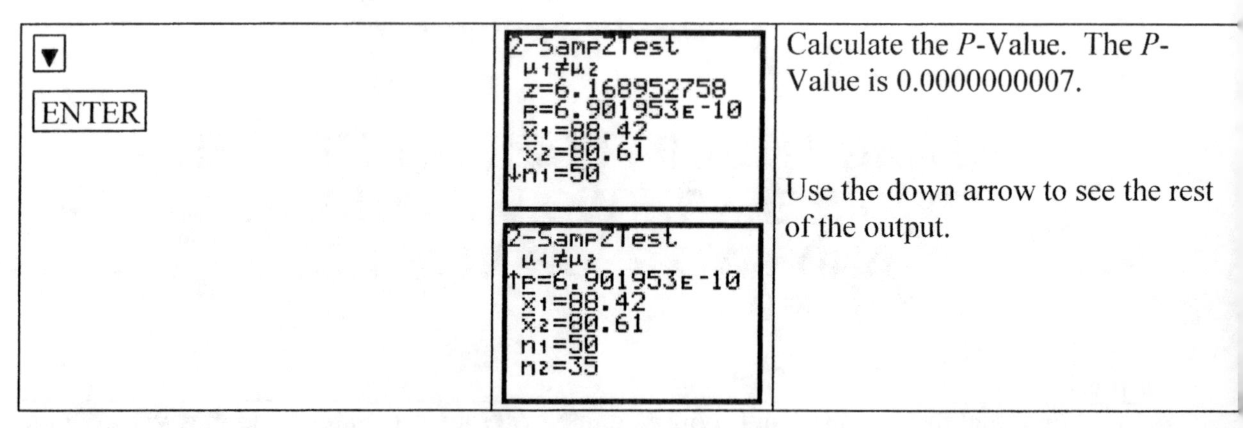

▼ ENTER	2-SampZTest μ1≠μ2 z=6.168952758 p=6.901953E-10 x̄1=88.42 x̄2=80.61 ↓n1=50 2-SampZTest μ1≠μ2 ↑p=6.901953E-10 x̄1=88.42 x̄2=80.61 n1=50 n2=35	Calculate the *P*-Value. The *P*-Value is 0.0000000007. Use the down arrow to see the rest of the output.

The null hypothesis is rejected since the *P*-Value is less than the level of significance. There is enough evidence to conclude that the room rate in New Orleans is not equal to the room rate in Phoenix.

Example 2

A discount store has two outlets for its merchandise. They wish to reduce their outlets to one. It is suspected that the outlet at Richland Center has significantly higher sales than the outlet at Dickson Center. A random sample of 35 days at Richland Center yields an average of $19,940 per day in sales with a standard deviation of $3,200 per day. Dickson Center had an average of $18,530 per day with a standard deviation of $2880 per day during the same 35 days.

(A) Calculate the test value to two decimal places to test the hypotheses:
$H_0 : \mu_R = \mu_D$ versus $H_1 : \mu_R > \mu_D$

(B) Suppose data was collected also for the next 16 days and combined with the existing data. The sample average for all 51 days for Richland Center was $19,830 per day with a standard deviation of $3420 per day. Dickson Center now has an average of $18,540 per day with a standard deviation of $2850 per day. Calculate the test value now.

(C) Test the hypotheses in Part (A) at the 0.05 level of significance.

(D) Test the hypotheses in Part (B) at the 0.05 level of significance.

(E) Are the results in Parts (C) and (D) the same? Why or why not?

Solution (A) & (C):

The information is entered into the calculator and a two-sample test on sample means is performed using 2-SampTTest... .

Keystrokes	*Screen Display*	*Comments*
CLEAR STAT ► ► 4 :2-SampTTest...	EDIT CALC TESTS 1:Z-Test... 2:T-Test... 3:2-SampZTest... 4:2-SampTTest... 5:1-PropZTest... 6:2-PropZTest... 7↓ZInterval...	Turn the calculator on and clear the Home Screen. Select two-sample *t* test from the statistics hypothesis tests menu.

▶ ENTER ▼ 19940 ENTER 3200 ENTER 35 ENTER 18530 ENTER 2880 ENTER 35 ENTER ▶ ▶ ENTER ▼ ENTER	2-SampTTest Inpt:Data Stats x̄1:19940 Sx1:3200 n1:35 x̄2:18530 Sx2:2880 ↓n2:35 2-SampTTest ↑n1:35 x̄2:18530 Sx2:2880 n2:35 μ1:≠μ2 <μ2 >μ2 Pooled:No Yes Calculate Draw	Enter the information. Select No for Pooled data since we cannot conclude that the standard deviations are equal. Select the greater than option.
▼ ENTER	2-SampTTest μ1>μ2 t=1.937598855 p=.0284362316 df=67.25886118 x̄1=19940 ↓x̄2=18530 2-SampTTest μ1>μ2 ↑x̄2=18530 Sx1=3200 Sx2=2880 n1=35 n2=35	Calculate the value of the test value and the *P*-Value. The test value is 1.94 and the *P*-Value is 0.03. Use the down arrow to see the rest of the output.

The null hypothesis is rejected since the *P*-Value is less than the level of significance. Hence, there is enough evidence to conclude that the outlet at Richland Center has significantly higher sales than the outlet at Dickson Center.

Solution (B) & (D):

The information is entered into the calculator and a two-sample test on sample means is performed using 2-SampTTest... .

Keystrokes	*Screen Display*	*Comments*
CLEAR		Turn the calculator on and clear the Home Screen.
STAT ▶ ▶ 4 :2-SampTTest...	EDIT CALC TESTS 1:Z-Test... 2:T-Test... 3:2-SampZTest... 4:2-SampTTest... 5:1-PropZTest... 6:2-PropZTest... 7↓ZInterval...	Select two-sample *t* test from the statistics hypothesis tests menu.

▶ ENTER ▼ 19830 ENTER 3420 ENTER 51 ENTER 18540 ENTER 2850 ENTER 51 ENTER	2-SampTTest Inpt:Data Stats x̄1:19830 Sx1:3420 n1:51 x̄2:18540 Sx2:2850 ↓n2:51	Enter the information.
▼ ▶ ENTER ▼ ENTER	2-SampTTest ↑n1:51 x̄2:18540 Sx2:2850 n2:51 μ1:≠μ2 <μ2 >μ2 Pooled:No Yes Calculate Draw	Select No for Pooled data since we cannot conclude that the standard deviations are equal. Select the greater than option.
▼ ENTER	2-SampTTest μ1>μ2 t=2.069355116 p=.0205879096 df=96.85059865 x̄1=19830 ↓x̄2=18540	Calculate the value of the critical value and the *P*-Value. The critical value is 2.07 and the *P*-Value is 0.02.

The null hypothesis is rejected since the *P*-Value is less than the level of significance. Hence, there is enough evidence to conclude that the outlet at Richland Center has significantly higher sales than the outlet at Dickson Center.

Solution (E):

The results are the same since the *P*-values for both situations are less than the level of significance.

Example 3

Graph the F distribution with 10 and 8 degrees of freedom.

Solution:

Keystrokes	*Screen Display*	*Comments*
CLEAR		Turn the calculator on and clear the Home Screen.
2nd DRAW 1 :ClrDraw ENTER	DRAW POINTS STO 1:ClrDraw 2:Line(3:Horizontal 4:Vertical 5:Tangent(6:DrawF 7↓Shade(	Clear the drawings.

[Y=] [CLEAR] [2nd] [DISTR] [8] :Fpdf(	Plot1 Plot2 Plot3 \Y1= \Y2= \Y3= \Y4= \Y5= \Y6= \Y7= DISTR DRAW 2↑normalcdf(3:invNorm(4:tpdf(5:tcdf(6:χ^2pdf(7:χ^2cdf(8:Fpdf(	Store the F distribution as Y1.
[X,T,θ,*n*] [,] [10] [,] [8] [)]	Plot1 Plot2 Plot3 \Y1=Fpdf(X,10,8) \Y2= \Y3= \Y4= \Y5= \Y6=	
[WINDOW] [0] [ENTER] [5] [ENTER] [1] [ENTER] [(-)] [.1] [ENTER] [.8] [ENTER] [1] [ENTER]	WINDOW Xmin=0 Xmax=5 Xscl=1 Ymin=-.1 Ymax=.8 Yscl=1 Xres=1	Enter the graph screen dimensions. You may need to experiment some to get the dimensions that work the best for the graph you wish to display.
[GRAPH]		Graph the distribution.

Example 4

Graph the F distribution with 10 and 8 degrees of freedom with the area between 1.25 and 2.58 shaded.

Solution:

Keystrokes	*Screen Display*	*Comments*
[CLEAR]		Turn the calculator on and clear the Home Screen.
[2nd] [DRAW] [1] :ClrDraw [ENTER] [Y=] [CLEAR]	DRAW POINTS STO 1:ClrDraw 2:Line(3:Horizontal 4:Vertical 5:Tangent(6:DrawF 7↓Shade(	Clear the drawings. Clear all the functions in the Y= list.

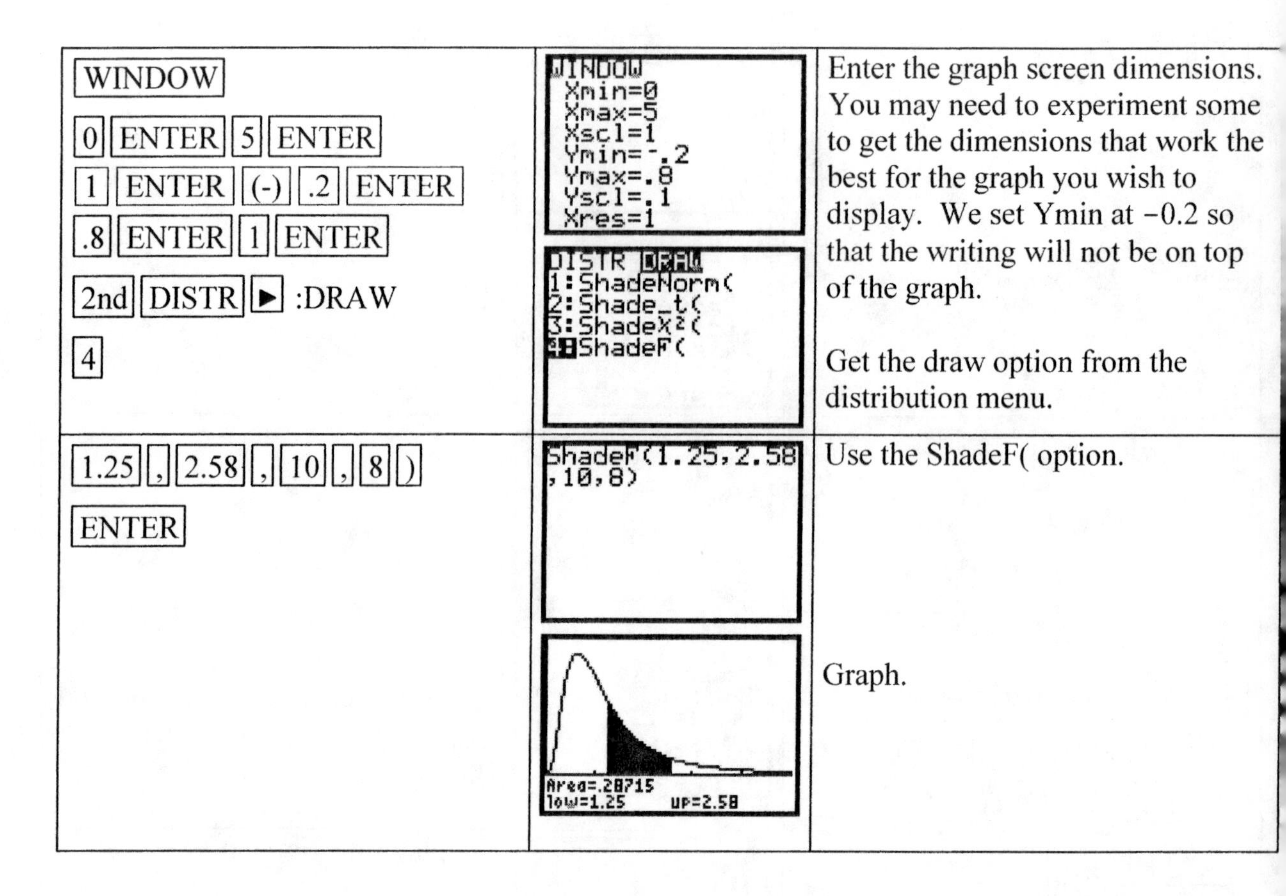

Keystrokes	Screen	Explanation
WINDOW 0 ENTER 5 ENTER 1 ENTER (-) .2 ENTER .8 ENTER 1 ENTER 2nd DISTR ▶ :DRAW 4	WINDOW Xmin=0 Xmax=5 Xscl=1 Ymin=-.2 Ymax=.8 Yscl=.1 Xres=1 DISTR DRAW 1:ShadeNorm(2:Shade_t(3:Shadeχ²(4:ShadeF(	Enter the graph screen dimensions. You may need to experiment some to get the dimensions that work the best for the graph you wish to display. We set Ymin at −0.2 so that the writing will not be on top of the graph. Get the draw option from the distribution menu.
1.25 , 2.58 , 10 , 8) ENTER	ShadeF(1.25,2.58,10,8) Area=.28715 low=1.25 up=2.58	Use the ShadeF(option. Graph.

Example 5

A medical researcher wishes to see whether the variance of the heart rates (in beats per minute) of smokers are different from the variance of heart rates of people who do not smoke. Two samples are selected and the data are:

Smokers	Nonsmokers
$n_S = 18$	$n_N = 26$
$s_S = 4$	$s_N = 6$

Test the hypothesis using 0.01 level of significance.

Solution:

The hypotheses to be tested are:

$$H_0 : \sigma_S^2 = \sigma_N^2 \quad \text{versus} \quad H_1 : \sigma_S^2 < \sigma_N^2$$

The information is entered into the calculator and a two-sample test on sample means is performed using 2-SampFTest... .

Keystrokes	*Screen Display*	*Comments*
[CLEAR]		Turn the calculator on and clear the Home Screen.
[STAT] [►] [►] [ALPHA] [D] :2-SampFTest...	EDIT CALC TESTS 0↑2-SampTInt... A:1-PropZInt... B:2-PropZInt... C:χ²-Test... D:2-SampFTest... E:LinRegTTest... F:ANOVA(	Select two-sample F test from the statistics hypothesis tests menu.
[►] [ENTER] [▼] [4] [ENTER] [18] [ENTER] [6] [ENTER] [26] [ENTER] [►] [ENTER]	2-SampFTest Inpt:Data Stats Sx1:4 n1:18 Sx2:6 n2:26 σ1:≠σ2 <σ2 >σ2 Calculate Draw	Enter the information. Select the less than option.
[▼] [►] :DRAW [ENTER]	F=.4444 p=.0438	It is a good idea to use the draw option on this test to check if the calculator has calculated the correct *P*-Value.
[STAT] [►] [►] [ALPHA] [D] :2-SampFTest... [▼] ...[▼] [ENTER]	2-SampFTest σ1<σ2 F=.4444444444 p=.043809078 Sx1=4 Sx2=6 ↓n1=18	Repeat the test using the calculate option to compare the outputs. Use the down arrow to see the rest of the output. The *P*-Value is 0.04.

The null hypothesis is not rejected since the *P*-Value is not less than the level of significance. There is not enough evidence to conclude that the variances of the heart rates of the smokers is less than the variance of the nonsmokers.

Example 6

Find the 95% confidence interval on the difference of two means for the data in Example 1. Round confidence interval limits to two decimal places.

Solution:

The two-sample *z* confidence interval is to be calculated.

Keystrokes	*Screen Display*	*Comments*
CLEAR		Turn the calculator on and clear the Home Screen.
STAT ▶ ▶ 9 :2-SampZInt...	EDIT CALC TESTS 4↑2-SampTTest... 5:1-PropZTest... 6:2-PropZTest... 7:ZInterval... 8:TInterval... 9:2-SampZInt... 0↓2-SampTInt...	Select two-sample z interval from the statistics hypothesis tests menu.
▶ ENTER ▼ 5.62 ENTER 5.83 ENTER 88.42 ENTER 50 ENTER 80.61 ENTER 35 ENTER 95 ENTER	2-SampZInt Inpt:Data Stats σ1:5.62 σ2:5.83 x̄1:88.42 n1:50 x̄2:80.61 ↓n2:35 2-SampZInt ↑σ2:5.83 x̄1:88.42 n1:50 x̄2:80.61 n2:35 C-Level:95 Calculate	Enter the information.
▼ ENTER	2-SampZInt (5.3287,10.291) x̄1=88.42 x̄2=80.61 n1=50 n2=35	

The confidence interval on the difference of two means is (5.33, 10.29). This means $P(5.33 < \mu_N - \mu_P < 10.29) = 0.95$.

Example 7

Find the 90% confidence interval on the difference of two means for the data in Example 2. Round confidence interval limits to two decimal places.

Solution:

The two-sample z confidence interval is to be calculated.

Keystrokes	*Screen Display*	*Comments*
CLEAR		Turn the calculator on and clear the Home Screen.
STAT ▶ ▶ 0 :2-SampTInt...	EDIT CALC TESTS 4↑2-SampTTest... 5:1-PropZTest... 6:2-PropZTest... 7:ZInterval... 8:TInterval... 9:2-SampZInt... 0↓2-SampTInt...	Select two-sample t interval from the statistics hypothesis tests menu.

<table>
<tr><td>[▶] [ENTER] [▼]
[19940] [ENTER] [3200] [ENTER]
[35] [ENTER] [18530] [ENTER]
[2880] [ENTER] [35] [ENTER]
[90] [ENTER] [ENTER]</td><td>2-SampTInt
Inpt:Data Stats
$\bar{x}$1:11940
Sx1:3200
n1:35
$\bar{x}$2:18530
Sx2:2880
↓n2:35

2-SampTInt
↑n1:35
$\bar{x}$2:18530
Sx2:2880
n2:35
C-Level:90
Pooled:No Yes
Calculate</td><td>Enter the information.

Do not pool the variances since we have no reason to believe they are equal.</td></tr>
<tr><td>[▼] [ENTER]</td><td>2-SampTInt
(196.31,2623.7)
df=67.25886118
$\bar{x}$1=19940
$\bar{x}$2=18530
Sx1=3200
↓Sx2=2880</td><td></td></tr>
</table>

The confidence interval on the difference of two means is (196.31, 2623.7). This means $P(196.31 < \mu_R - \mu_D < 2623.70) = 0.95$.

Example 8

Test the hypothesis that there is no difference in population means based on the following sample of paired data. Use 0.05 as the level of significance.

Set A	33	35	28	29	32	34	30	34
Set B	27	29	36	34	30	29	28	24

Solution:

Find the difference of the two sets by entering the data of Set A into L1, entering the data of set B into L2, and storing the difference of L1-L2 in L3. Then perform a *t* test on the data in L3.

Keystrokes	*Screen Display*	*Comments*
[CLEAR]		Turn the calculator on and clear the Home Screen.
[STAT] [4] :ClrList [2nd] [L1] [,] [2nd] [L2] [,] [2nd] [L3] [ENTER]	EDIT CALC TESTS 1:Edit... 2:SortA(3:SortD(4:ClrList 5:SetUpEditor ClrList L1,L2,L3 Done	Clear lists L1, L2 and L3.

Keystrokes	Screen	Explanation
STAT 1 :Edit 33 ENTER 35 ENTER etc. ► 27 ENTER 29 ENTER etc.	L1: 28 29 32 34 30 34 L2: 36 34 30 29 28 24 L2(9) =	Enter the data in L1. Enter the data in L2.
2nd QUIT 2nd L1 - 2nd L2 STO► 2nd L3 ENTER	ClrList L1,L2 Done L1-L2→L3 {6 6 -8 -5 2 5 ...	Return to the Home Screen. Store the difference between L1 and L2 in L3. The rest of the list can be read using the right arrow keys or using Edit from the STAT menu.
STAT ► ► 2 :T-TEST	EDIT CALC TESTS 1:Z-Test... 2:T-Test... 3:2-SampZTest... 4:2-SampTTest... 5:1-PropZTest... 6:2-PropZTest... 7↓ZInterval...	Get the t test from the statistics hypothesis test menu.
ENTER ◄ ENTER ▼ 0 ENTER 2nd L3 ENTER ALPHA 1 ENTER ◄ ENTER	T-Test Inpt:Data Stats μ0:0 List:L3 Freq:1 μ:≠μ0 <μ0 >μ0 Calculate Draw	Select the Data option. Enter the other information. Select the not-equal-to option.
▼ ENTER	T-Test μ≠0 t=1.057517468 p=.3253987995 x̄=2.25 Sx=6.017830649 n=8	Calculate. The *P*-Value is 0.33

The null hypothesis is not rejected since the *P*-Value is not less than the level of significance. There is not enough evidence to conclude that the difference in population means of these dependent samples is different from zero.

Example 9

A sample of 50 randomly selected men with high triglyceride levels consumed two tablespoons of oat bran daily for six weeks. After six weeks, 60% of the men had lowered their triglyceride level. A sample of 80 men consumed two tablespoons of wheat bran for six weeks. After six weeks, 25% had lower triglyceride levels.

(A) Is there a significant difference in the two proportions at the 0.01 significance level?

(B) Find a 99% confidence interval on the difference of the population proportions.

Solution (A):

Keystrokes	*Screen Display*	*Comments*
CLEAR		Turn the calculator on and clear the Home Screen.
STAT ► ► 6 :2-PropZTest...	EDIT CALC TESTS 1:Z-Test... 2:T-Test... 3:2-SampZTest... 4:2-SampTTest... 5:1-PropZTest... 6:2-PropZTest... 7↓ZInterval...	Get the two proportion z test from the hypothesis test menu.
30 ENTER 50 ENTER 20 ENTER 80 ENTER ◄ ENTER	2-PropZTest x1:30 n1:50 x2:20 n2:80 p1:≠p2 <p2 >p2 Calculate Draw	Enter the information. Calculate x1 as 0.60(50). This can be entered in-to the calculator as: .60 (50) . The calculator will accept this expression. However if the result is a decimal then multiply all the numbers x1, n1, x2, n2 decimal by the same power of 10 to make the decimal a whole number. Select the not-equal-to option.
▼ ENTER	2-PropZTest p1≠p2 z=3.990613988 p=6.5932804E-5 p̂1=.6 p̂2=.25 ↓p̂=.3846153846 2-PropZTest p1≠p2 ↑p̂1=.6 p̂2=.25 p̂=.3846153846 n1=50 n2=80	Calculate. The P-Value is .000066. Use the down arrow key to see the rest of the information.

The null hypothesis is rejected since the P-Value of 0.00007 is less than the level of significance. There is enough evidence to reject the claim that there is no difference in the proportions of men who had lowered triglyceride levels consuming the different types of bran.

Solution (B):

Keystrokes	*Screen Display*	*Comments*
CLEAR		Turn the calculator on and clear the Home Screen.

STAT ▶ ▶ ALPHA B :2-PropZInt...	EDIT CALC TESTS 6↑2-PropZTest... 7:ZInterval... 8:TInterval... 9:2-SampZInt... 0:2-SampTInt... A:1-PropZInt... B:2-PropZInt...	Get the two proportion z interval from the hypothesis test menu.
30 ENTER 50 ENTER 20 ENTER 80 ENTER 99 ENTER	2-PropZInt x1:30 n1:50 x2:20 n2:80 C-Level:99 Calculate	Enter the information.
ENTER	2-PropZInt (.13229,.56771) p̂1=.6 p̂2=.25 n1=50 n2=80	Calculate the confidence interval.

The confidence interval for the difference between two proportions can be written:

$$P(0.132 < p_O - p_W < 0.568) = 0.99.$$

EXERCISE SET 9

1. A researcher wishes to see if there is a difference in the cholesterol levels of two groups of men. A random sample of 30 men between the ages of 25 and 40 is selected and tested. The average level is 223. A second sample of 25 men between the ages of 41 and 56 is selected and tested. The average of this group is 229. The population standard deviation for both groups is 6.

   (A) At the 0.01 level of significance, is there a difference in the cholesterol levels between the two groups?

   (B) Find the 99% confidence interval for the difference of the two means.

2. Repeat Problem #1 using $\alpha = 0.05$ and a 95% confidence interval. Use two decimal place accuracy.

3. Two groups of drivers are surveyed to see how many miles per week they drive for pleasure trips. The data are shown below.

Single drivers				
106	110	115	121	132
119	97	118	122	135
110	117	116	138	142
115	114	103	98	99
108	117	152	147	117
154	86	115	116	104
107	133	138	142	140

Married drivers				
97	103	138	102	115
133	120	119	136	96
139	108	117	145	114
140	136	113	113	150
101	114	116	113	135
115	109	147	106	88
113	119	99	108	105

(A) At 0.01 significance level can it be concluded that single drivers do more driving for pleasure trips on average than married drivers?

(B) Find the 99% confidence interval on the difference between two means. Use two decimal place accuracy.

4. Repeat Problem #3 using $\alpha = 0.05$ and a 95% confidence interval.

5. Graph the F distribution with 15 and 18 degrees of freedom.

6. Graph the F distribution with 20 and 12 degrees of freedom.

7. Graph the F distribution with 13 and 20 degrees of freedom with the area between 0.4 and 1.5 shaded.

8. Graph the F distribution with 19 and 8 degrees of freedom with the area between 0.7 and 1.9 shaded.

9. An educator wishes to compare the standard deviations of the amount of money spend per pupil in two states. The data are given below. At the 0.01 level of significance is there a difference in the standard deviations of the amounts the states spend per pupil?

State 1	**State 2**
$n_1 = 20$	$n_2 = 23$
$s_1 = \$28.12$	$s_2 = \$26.39$

10. Repeat Problem 9 where additional sampling yielded:

State 1	**State 2**
$n_1 = 40$	$n_2 = 53$
$s_1 = \$27.12$	$s_2 = \$36.39$

11. In an effort to improve the vocabulary of 10 students, a teacher provided a weekly one-hour tutoring session for them. A pretest is given before the sessions and a posttest is given afterward. The results are shown in the table. At the 0.05 significance level, can the teacher conclude that the tutoring sessions helped to improve the students' vocabulary?

Student	**1**	**2**	**3**	**4**	**5**	**6**	**7**	**8**	**9**	**10**
Pretest	83	76	92	64	82	68	70	71	72	63
Posttest	88	82	100	72	81	75	79	68	81	70

12. Repeat Problem 11 with two additional students:

Student	**11**	**12**
Pretest	78	90
Posttest	62	80

13. In a recent survey of 45 apartment residents, 28 had phone answering machines. In a survey of 55 homeowners, 39 had phone answering machines.

    (A) At the 0.01 level of significance, test the claim that the proportions are equal.

    (B) Find the 99% confidence interval for the difference of the two proportions.

14. In a sample of 80 workers from a factory in City A, it was found that 4% were unable to read. In a sample of 40 workers in city B, 9% were unable to read.

    (A) Can it be concluded that there is a difference in the proportions of nonreaders in the two cities using a 0.05 level of significance?

    (B) Find the 95% confidence interval for the difference of the two proportions.

CHAPTER 10

CORRELATION AND REGRESSION

Example 1

A study of age and systolic blood pressure of six randomly selected subjects yielded the following data:

Subject	Age, x	Pressure, y
A	43	128
B	48	120
C	56	135
D	61	143
E	67	141
F	70	152

(A) Construct a scatter plot using age as the independent variable.

(B) Find the linear correlation coefficient.

(C) Find the equation of the line of best fit.

(D) Graph the scatter plot and the regression line on the same set of coordinate axes.

(E) Test the significance of a positive correlation between the variables.

(F) Find the coefficient of determination.

Round all numbers to two decimal places.

Solution (A):

Enter the information into the calculator in list L1 and list L2.

Keystrokes	*Screen Display*	*Comments*
[CLEAR]		Turn the calculator on and clear the Home Screen.
[STAT] [4] :ClrList [2nd] [L1] [,] [2nd] [L2] [ENTER]	ClrList L1,L2 Done	Clear the lists.

STAT 1 :Edit 43 ENTER 48 ENTER etc. ▶ 128 ENTER 120 ENTER 135 ENTER etc.	L1 L2 L3 2 43 128 ------ 48 120 56 135 61 143 67 141 70 152 ------ L2(7) =	Enter the data into list L1 and list L2. Use the arrow keys to select the appropriate list.
WINDOW 40 ENTER 75 ENTER 5 ENTER 115 ENTER 150 ENTER 5 ENTER 1 ENTER 2nd QUIT	WINDOW Xmin=40 Xmax=75 Xscl=5 Ymin=115 Ymax=150 Yscl=5 Xres=1	Set the graph screen dimensions. Xmin=40 since 43 is the least age, Xmax=75 since 70 is the greatest age, etc.
2nd STAT PLOT 1 :Plot1...Off ENTER ▼ ENTER ▼ 2nd L1 ENTER 2nd L2 ENTER ENTER	STAT PLOTS 1:Plot1...Off L1 L3 2:Plot2...Off L1 L2 3:Plot3...Off L1 L2 4↓PlotsOff Plot1 Plot2 Plot3 On Off Type: Xlist:L1 Ylist:L2 Mark:	Turn Plot 1 on and set the kind of graph desired.
GRAPH		Graph the scatter plot.

Solution (B), (C) & (F):

Keystrokes	*Screen Display*	*Comments*
CLEAR		Turn the calculator on and clear the Home Screen.

STAT ► 4 :LinReg(ax+b) 2nd L1 , 2nd L2 ENTER	EDIT CALC TESTS 1:1-Var Stats 2:2-Var Stats 3:Med-Med 4:LinReg(ax+b) 5:QuadReg 6:CubicReg 7↓QuartReg	Select the linear regression option from the calculation menu.
	LinReg(ax+b) L1, L2	An alternative method would be to use Option 8:LinReg(a+bx) which will also calculate the coefficients for the line. The form of the algebraic expression for the line is different.
	LinReg y=ax+b a=.964381122 b=81.04808549	We see the coefficients of the regression equation but do not have the correlation coefficient displayed.
2nd CATALOG ▼ ... ▼ ENTER ENTER	a=.964381122 b=81.04808549 DiagnosticOn Done	Get the catalog list and arrow down 41 lines on the TI-83 and 45 lines on the TI-83+ to DiagnosticOn. Press ENTER twice to turn the diagnostics on.
STAT ► 4 :LinReg(ax+b) 2nd L1 , 2nd L2 ENTER	LinReg y=ax+b a=.964381122 b=81.04808549 r²=.8040221364 r=.8966728145	Repeat the linear regression calculation. Now the correlation coefficient and the coefficient of determination appears on the screen.

The Pearson product moment correlation coefficient is 0.90. The line of best fit is $y' = 0.96x + 81.05$. The coefficient of determination is 0.804. This means that 80.4% of the variation of the dependent variable is explained by the regression line and the independent variable.

Solution (D):

The scatter plot was graphed in Part (A). Now store the equation of the line in the Y= list and graph.

Keystrokes	*Screen Display*	*Comments*
CLEAR		Turn the calculator on and clear the Home Screen.

Keys	Screen	Explanation
Y= CLEAR VARS 5 :Statistics	VARS Y-VARS 1:Window... 2:Zoom... 3:GDB... 4:Picture... 5:Statistics... 6:Table... 7:String... XY Σ EQ TEST PTS 1:n 2:x̄ 3:Sx 4:σx 5:ȳ 6:Sy 7↓σy	Get the Y= list, then the VARS menu and select the statistics option.
▶ ▶ 1 :RegEQ	XY Σ EQ TEST PTS 1:RegEQ 2:a 3:b 4:c 5:d 6:e 7↓r Plot1 Plot2 Plot3 \Y1=.96438112199466X+81.048085485307 \Y2= \Y3= \Y4= \Y5=	Select the regression equation on the equation menu. The equation will be placed in the Y= list.
GRAPH		Graph the scatter plot and line.
TRACE	P1:L1,L2 X=43 Y=128	Use the trace feature of the calculator to trace the line or the points. Use the up or down arrow to select either the points or the line (see upper left corner of the screen). Use the right arrow to move from point to point or along the line.
	Y1=.96438112199466X+81.0_ X=57.5 Y=136.5	

Another way to get the regression equation into the Y= list is to enter 1's as list L3. Then use the command LinReg(ax+b) L1, L2, L3, Y1.

STAT ▶ 4 :LinReg(ax+b) 2nd L1 , 2nd L2 , 2nd L3 , VARS
▶ :Function... 1 :Y1 ENTER .

Press Y= to see that the regression equation has been placed in the list as Y1.

Solution (E):

Keystrokes	*Screen Display*	*Comments*
CLEAR		Turn the calculator on and clear the Home Screen.
STAT ▶ ▶ ALPHA E :LinRegTTest.. 2nd L1 ENTER 2nd L2 ENTER 1 ENTER ▶ ▶ ENTER ▼ VARS ▶ 1 :Function 1 :Y1 ENTER	EDIT CALC TESTS 9↑2-SampZInt… 0:2-SampTInt… A:1-PropZInt… B:2-PropZInt… C:χ²-Test… D:2-SampFTest… E:LinRegTTest… LinRegTTest Xlist:L1 Ylist:L2 Freq:1 β & ρ:≠0 <0 >0 RegEQ:Y1 Calculate	Select the linear regression *t* test from the hypothesis test menu. Enter the lists L1 and L2 where the data is stored. The frequency is 1. This is a greater than 0 alternative. The equation is stored in Y1. Get this from the variables function list.
ENTER	LinRegTTest y=a+bx β>0 and ρ>0 t=4.050983638 p=.0077315871 df=4 ↓a=81.04808549 LinRegTTest y=a+bx β>0 and ρ>0 ↑b=.964381122 s=5.641090817 r²=.8040221364 r=.8966728145	The *P*-Value of this test is 0.0077315871. This is less than the level of significance.

Reject the hypothesis that there is no correlation between the two variables.
There is enough evidence to conclude that there is a positive correlation between the variables.

Example 2

A study yields the following data:

Before	4	6	5	9	12	3	4
After	15	37	24	83	126	12	20

(A) Find the correlation coefficient or coefficient of determination for the following models:

$y' = ax^2 + bx + c$ $\quad y' = ax^3 + bx^2 + cx + d$ $\quad y' = ax^4 + bx^3 + cx^2 + dx + e$

$y' = a + b \ln x$ $\quad y' = a(b^x)$ $\quad y' = ax^b$ $\quad y' = \dfrac{c}{1 + ae^{-bx}}$

(B) Write the regression equation for each of these models.

(C) Draw the scatter plot and graph the regression equation for each model.

Use two decimal places in all coefficients.

Solution (A), (B) & (C):

Enter the information into the calculator in list L1 and list L2.

Keystrokes	*Screen Display*	*Comments*
[CLEAR]		Turn the calculator on and clear the Home Screen.
[STAT] [4] :ClrList [2nd] [L1] [,] [2nd] [L2] [,] [2nd] [L3] [ENTER]	ClrList L1,L2,L3 Done	Clear the lists.
[STAT] [1] :Edit [4] [ENTER] [6] [ENTER] etc. [▶] [15] [24] [ENTER] etc.	L1: 6, 5, 9, 12, 3, 4 L2: 37, 24, 83, 126, 12, 20 L3: 1, 1, 1, 1, 1, 1 L3(8) =	Enter the data into list L1 and list L2. Enter 1's into L3. Use the arrow keys to move from one list to another.
[WINDOW] [0] [ENTER] [15] [ENTER] [1] [ENTER] [0] [ENTER] [140] [ENTER] [5] [ENTER] [1] [ENTER] [2nd] [QUIT]	WINDOW Xmin=0 Xmax=15 Xscl=1 Ymin=0 Ymax=140 Yscl=5 Xres=1	Set the graph screen dimensions.
[2nd] [STAT PLOT] [1] :Plot1...Off [ENTER] [▼] [ENTER] [▼] [2nd] [L1] [ENTER] [2nd] [L2] [ENTER] [ENTER]	STAT PLOTS 1:Plot1...Off L1 L3 2:Plot2...Off L1 L2 3:Plot3...Off L1 L2 4↓PlotsOff Plot1 Plot2 Plot3 On Off Type: Xlist:L1 Ylist:L2 Mark: ▫ + ·	Turn Plot 1 on and set the kind of graph desired. Use the arrow keys to select the appropriate option.

Quadratic Model

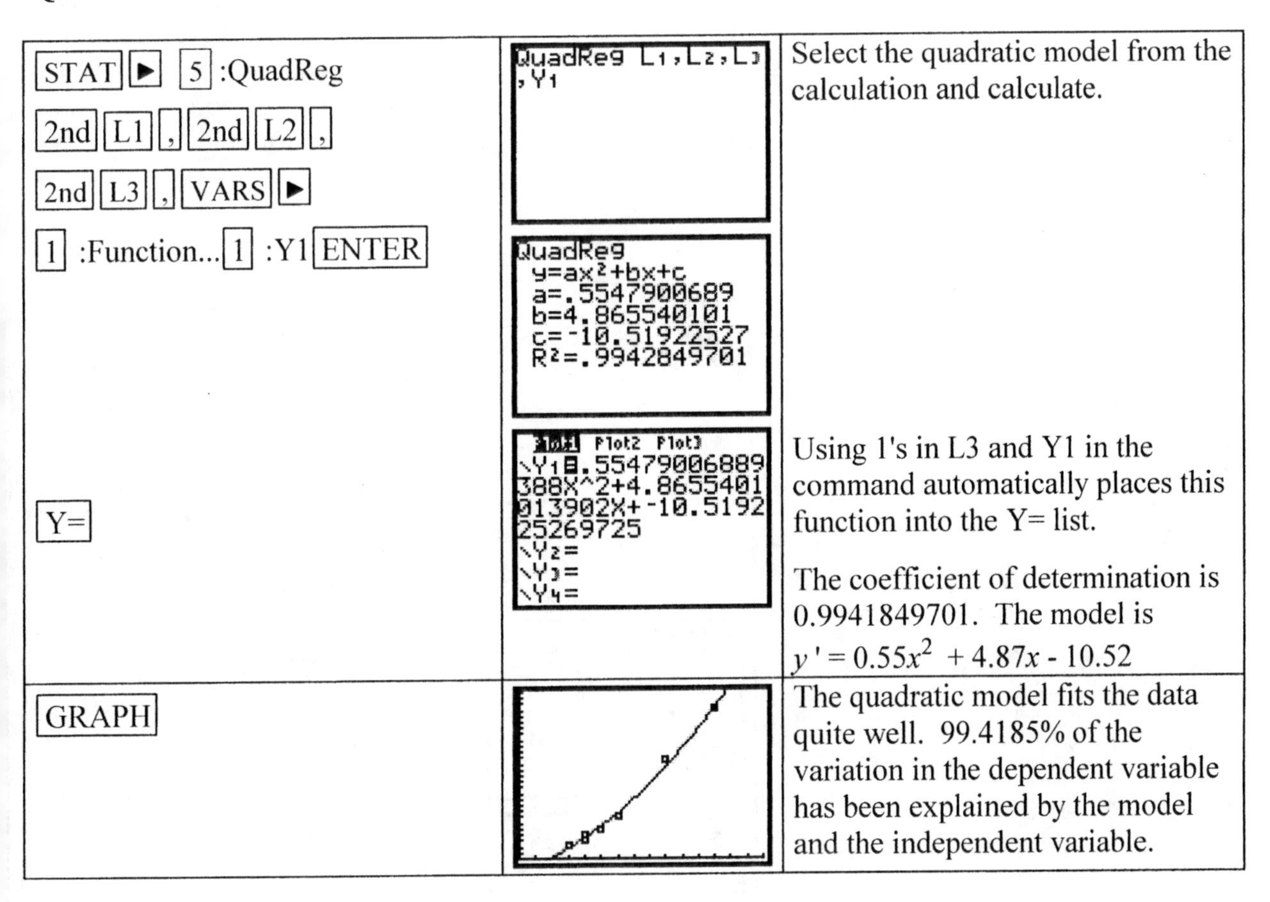

STAT ► 5 :QuadReg 2nd L1 , 2nd L2 , 2nd L3 , VARS ► 1 :Function... 1 :Y1 ENTER	QuadReg L1,L2,L3 ,Y1 QuadReg y=ax²+bx+c a=.5547900689 b=4.865540101 c=-10.51922527 R²=.9942849701	Select the quadratic model from the calculation and calculate.
Y=	Plot1 Plot2 Plot3 \Y1=.55479006889 388X^2+4.8655401 013902X+-10.5192 25269725 \Y2= \Y3= \Y4=	Using 1's in L3 and Y1 in the command automatically places this function into the Y= list. The coefficient of determination is 0.9941849701. The model is $y' = 0.55x^2 + 4.87x - 10.52$
GRAPH		The quadratic model fits the data quite well. 99.4185% of the variation in the dependent variable has been explained by the model and the independent variable.

Cubic Model

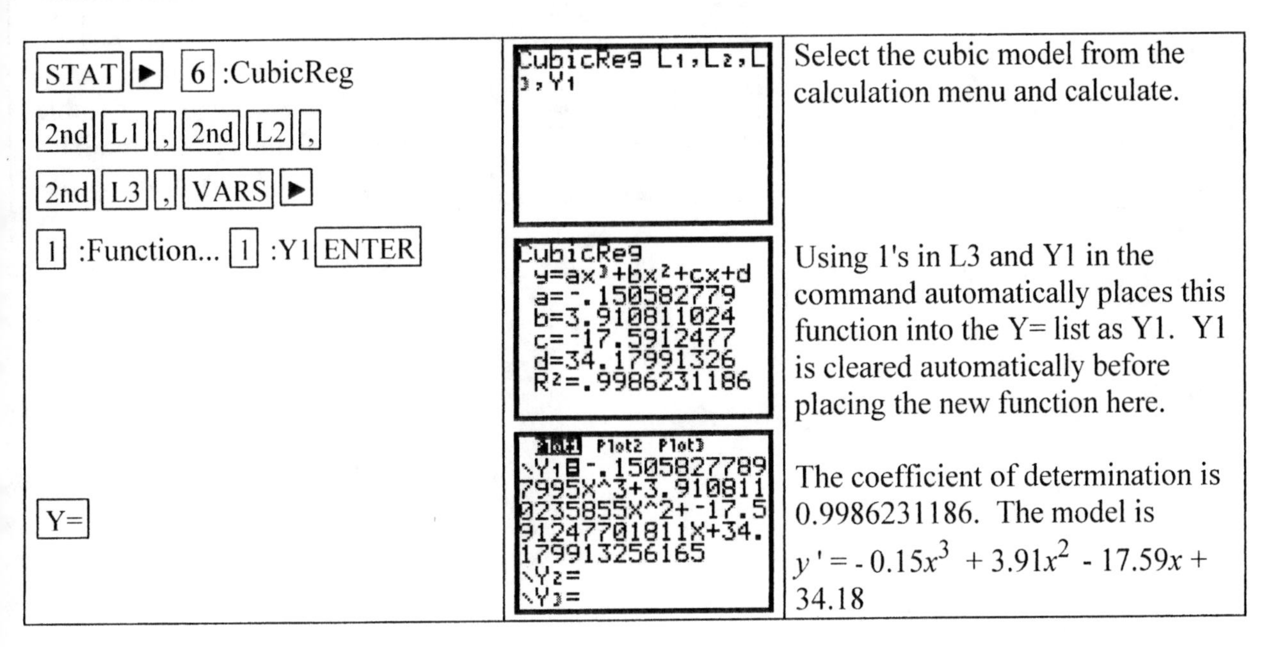

STAT ► 6 :CubicReg 2nd L1 , 2nd L2 , 2nd L3 , VARS ► 1 :Function... 1 :Y1 ENTER	CubicReg L1,L2,L 3,Y1	Select the cubic model from the calculation menu and calculate.
	CubicReg y=ax³+bx²+cx+d a=-.150582779 b=3.910811024 c=-17.5912477 d=34.17991326 R²=.9986231186	Using 1's in L3 and Y1 in the command automatically places this function into the Y= list as Y1. Y1 is cleared automatically before placing the new function here.
Y=	Plot1 Plot2 Plot3 \Y1=-.1505827789 7995X^3+3.910811 0235855X^2+-17.5 91247701811X+34. 179913256165 \Y2= \Y3=	The coefficient of determination is 0.9986231186. The model is $y' = -0.15x^3 + 3.91x^2 - 17.59x + 34.18$

GRAPH		The cubic model fits the data quite well. 99.8623% of the variation in the dependent variable has been explained by the model and the independent variable.

Quartic Model

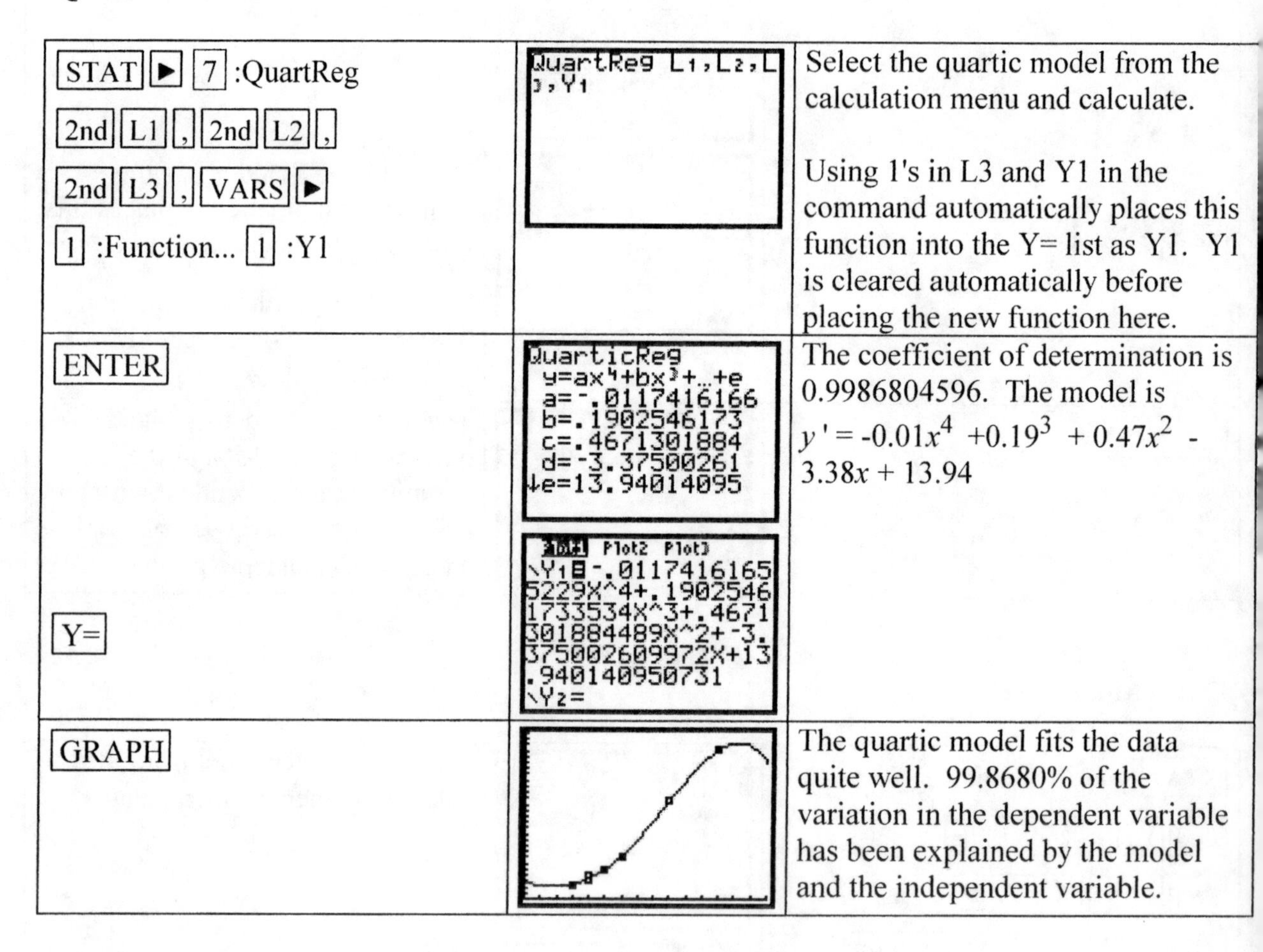

STAT ▶ 7 :QuartReg 2nd L1 , 2nd L2 , 2nd L3 , VARS ▶ 1 :Function... 1 :Y1		Select the quartic model from the calculation menu and calculate. Using 1's in L3 and Y1 in the command automatically places this function into the Y= list as Y1. Y1 is cleared automatically before placing the new function here.
ENTER Y=		The coefficient of determination is 0.9986804596. The model is $y' = -0.01x^4 + 0.19^3 + 0.47x^2 - 3.38x + 13.94$
GRAPH		The quartic model fits the data quite well. 99.8680% of the variation in the dependent variable has been explained by the model and the independent variable.

Logarithmic Model

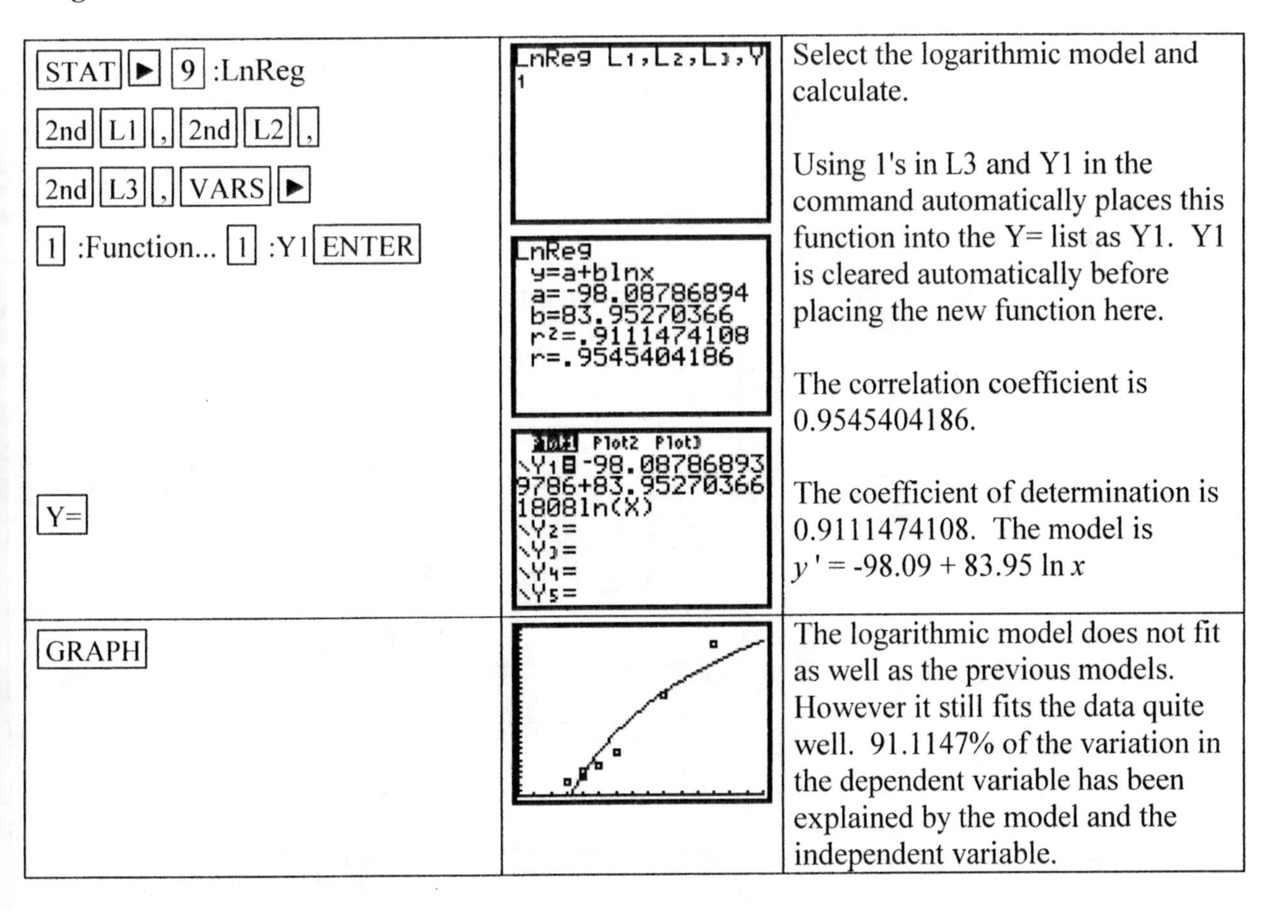

Keystrokes	Screen	Explanation
STAT ► 9 :LnReg 2nd L1 , 2nd L2 , 2nd L3 , VARS ► 1 :Function... 1 :Y1 ENTER	LnReg L1,L2,L3,Y1 LnReg y=a+blnx a=-98.08786894 b=83.95270366 r²=.9111474108 r=.9545404186	Select the logarithmic model and calculate. Using 1's in L3 and Y1 in the command automatically places this function into the Y= list as Y1. Y1 is cleared automatically before placing the new function here. The correlation coefficient is 0.9545404186.
Y=	Plot1 Plot2 Plot3 \Y1■-98.087868939786+83.952703661808ln(X) \Y2= \Y3= \Y4= \Y5=	The coefficient of determination is 0.9111474108. The model is $y' = -98.09 + 83.95 \ln x$
GRAPH		The logarithmic model does not fit as well as the previous models. However it still fits the data quite well. 91.1147% of the variation in the dependent variable has been explained by the model and the independent variable.

Exponential Model

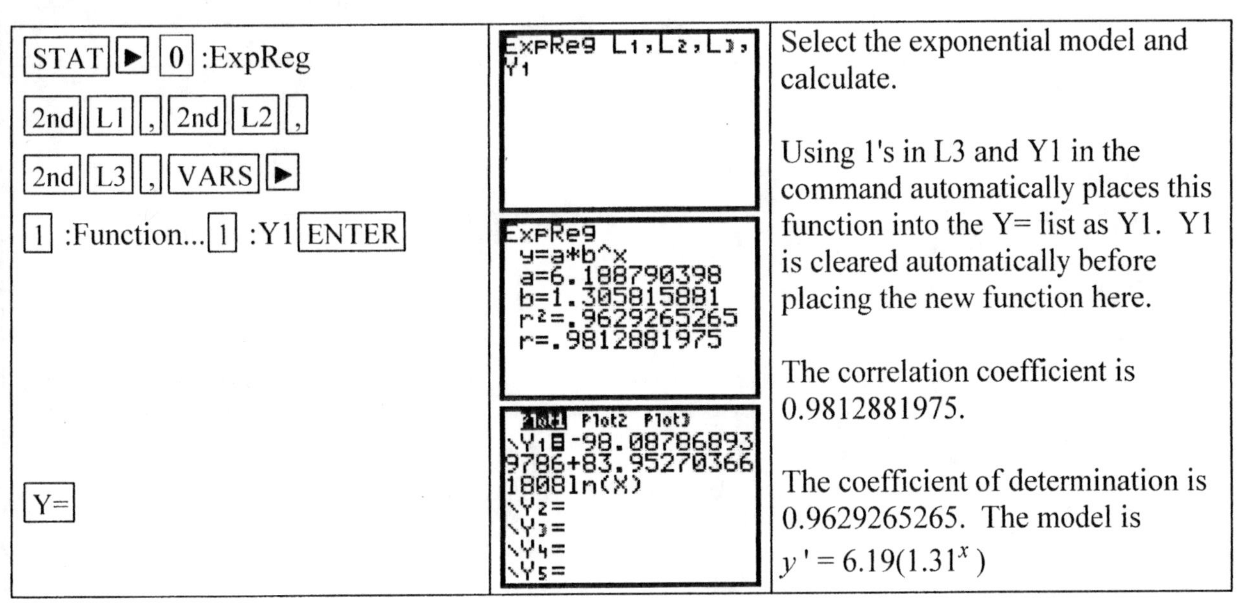

Keystrokes	Screen	Explanation
STAT ► 0 :ExpReg 2nd L1 , 2nd L2 , 2nd L3 , VARS ► 1 :Function... 1 :Y1 ENTER	ExpReg L1,L2,L3,Y1 ExpReg y=a*b^x a=6.188790398 b=1.305815881 r²=.9629265265 r=.9812881975	Select the exponential model and calculate. Using 1's in L3 and Y1 in the command automatically places this function into the Y= list as Y1. Y1 is cleared automatically before placing the new function here. The correlation coefficient is 0.9812881975.
Y=	Plot1 Plot2 Plot3 \Y1■-98.087868939786+83.952703661808ln(X) \Y2= \Y3= \Y4= \Y5=	The coefficient of determination is 0.9629265265. The model is $y' = 6.19(1.31^x)$

GRAPH		96.29% of the variation in the dependent variable has been explained by the model and the independent variable.

Power Model

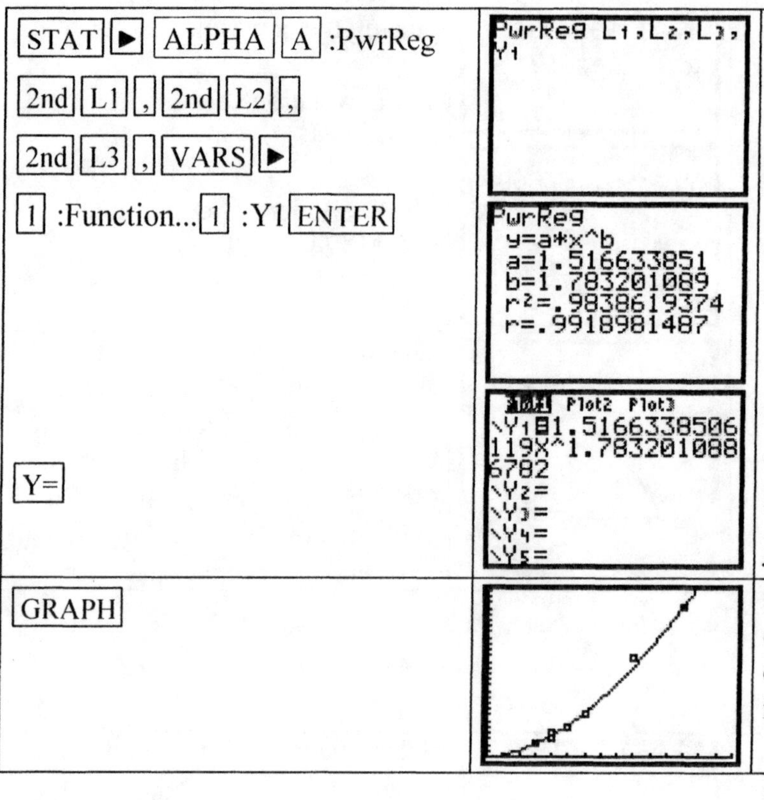

STAT ► ALPHA A :PwrReg 2nd L1 , 2nd L2 , 2nd L3 , VARS ► 1 :Function... 1 :Y1 ENTER Y=		Select the power model and calculate. Using 1's in L3 and Y1 in the command automatically places this function into the Y= list as Y1. Y1 is cleared automatically before placing the new function here. The correlation coefficient is 0.9918981487 The coefficient of determination is 0.9838619374. The model is $y' = 1.52(x^{1.78})$
GRAPH		98.39% of the variation in the dependent variable has been explained by the model and the independent variable.

Logistic Model

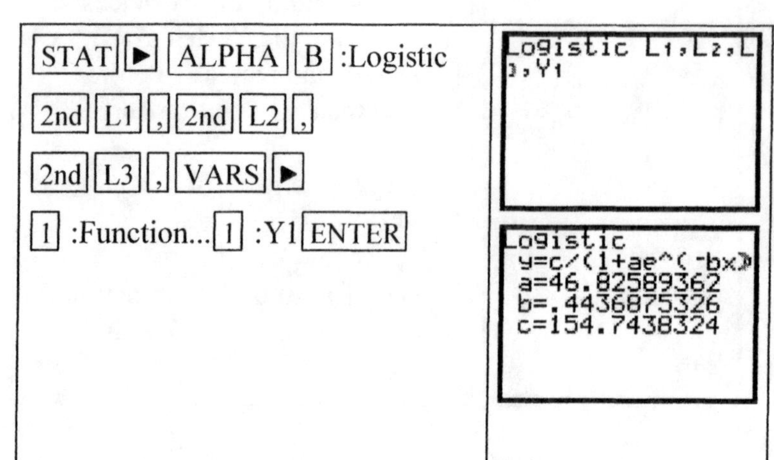

STAT ► ALPHA B :Logistic 2nd L1 , 2nd L2 , 2nd L3 , VARS ► 1 :Function... 1 :Y1 ENTER		Select the logistic model and calculate. Using 1's in L3 and Y1 in the command automatically places this function into the Y= list as Y1. Y1 is cleared automatically before placing the new function here.

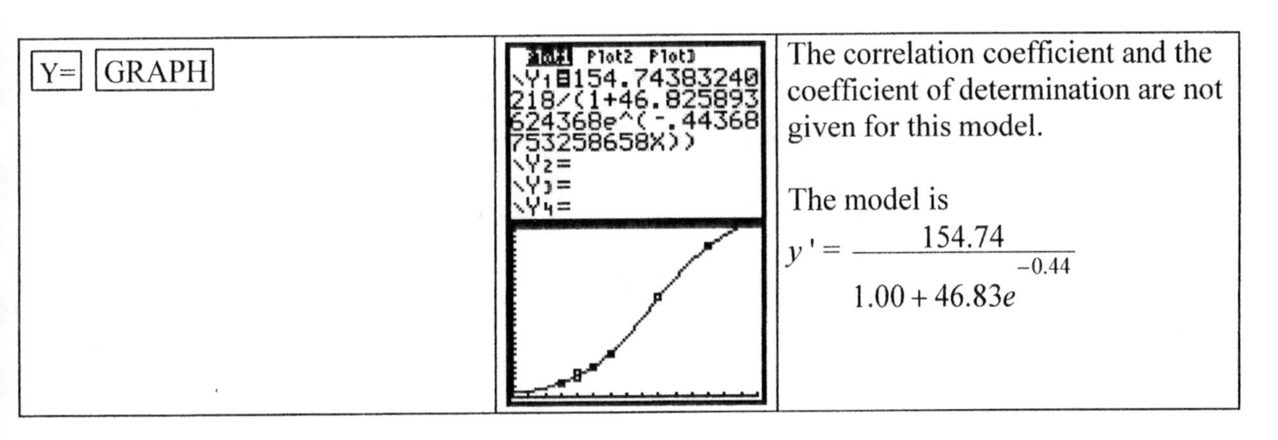

Y= GRAPH

The correlation coefficient and the coefficient of determination are not given for this model.

The model is

$$y' = \frac{154.74}{1.00 + 46.83e^{-0.44}}$$

EXERCISE SET 10

1. A study was conducted with vegetarians to see whether the number of grams of protein each ate per day was related to diastolic blood pressure.

Person	A	B	C	D	E	F	G	H
Grams, x	4.0	6.5	8.0	5.5	9.0	8.0	10.0	10.5
Pressure, y	73	79	83	80	87	85	93	98

(A) Construct a scatter plot.

(B) Find the linear correlation coefficient.

(C) Find the equation of the line of best fit.

(D) Graph the scatter plot and the regression line on the same set of coordinate axes.

(E) Test the significance of a positive correlation between the variables.

(F) Find the coefficient of determination.

2. A sample of eight students was taken at a small community college. Their grade point averages and the number of hours they worked per week were recorded. It is thought that the number of hours worked influences the grade point average.

Student	A	B	C	D	E	F	G	H
Hours worked	10	15	13	18	14	13	20	19
Grade Point Average	3.2	2.7	2.3	2.5	3.8	3.4	2.2	1.9

(A) Construct a scatter plot.

(B) Find the linear correlation coefficient.

(C) Find the equation of the line of best fit.

(D) Graph the scatter plot and the regression line on the same set of coordinate axes.

(E) Test the significance of a positive correlation between the variables.

(F) Find the coefficient of determination.

3. Use the data in Problem 1.

(A) Find the correlation coefficient and/or coefficient of determination for the following models:

$y' = ax^2 + bx + c$ $\quad y' = ax^3 + bx^2 + cx + d$ $\quad y' = ax^4 + bx^3 + cx^2 + dx + e$

$y' = a + b \ln x$ $\quad y' = a(b^x)$ $\quad y' = a x^b$

(B) Write the regression equation for each of these models.

(C) Draw the scatter plot and graph the regression equation for each model.

Use two decimal places in all coefficients.

4. Use the data in Problem 2.

(A) Find the correlation coefficient and/or coefficient of determination for the following models:

$y' = ax^2 + bx + c$ $\quad y' = ax^3 + bx^2 + cx + d$ $\quad y' = ax^4 + bx^3 + cx^2 + dx + e$

$y' = a + b \ln x$ $\quad y' = a(b^x)$ $\quad y' = a x^b$ $\quad y' = \dfrac{c}{1 + ae^{-bx}}$

(B) Write the regression equation for each of these models.

(C) Draw the scatter plot and graph the regression equation for each model.

Use two decimal places in all coefficients.

CHAPTER 11

OTHER CHI-SQUARE TESTS

Example 1

A market analyst wished to see whether consumers have a preference among five flavors of a new fruit soda. A sample of 100 people provided the following data:

Cherry	Strawberry	Orange	Lime	Grape
32	28	16	14	10

(A) Calculate the test value for testing:

H_0: Consumers show no preference for flavors of the fruit soda.

H_1: Consumers show a preference

(B) Graph the observed values and expected values for soda flavors.

Solution (A):

Keystrokes	*Screen Display*	*Comments*
[CLEAR]		Turn the calculator on and clear the Home Screen.
[STAT] [4] :ClrList [2nd] [L1] [,] [2nd] [L2] [,] [2nd] [L3] [,] [2nd] [L4] [ENTER]		Clear the lists.
[STAT] [1] :Edit [32] [ENTER] [28] [ENTER] etc. [▶] [20] [ENTER] [20] [ENTER] etc. [▶] [▶] [1] [ENTER] [2] [ENTER] etc.	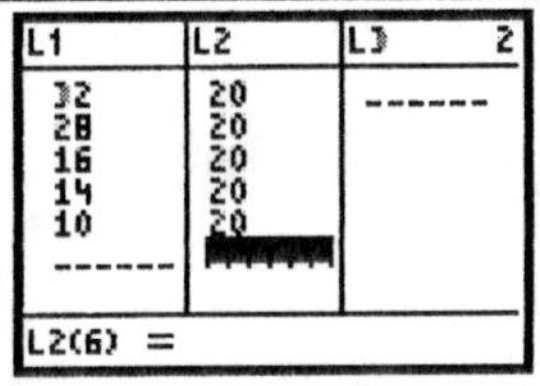 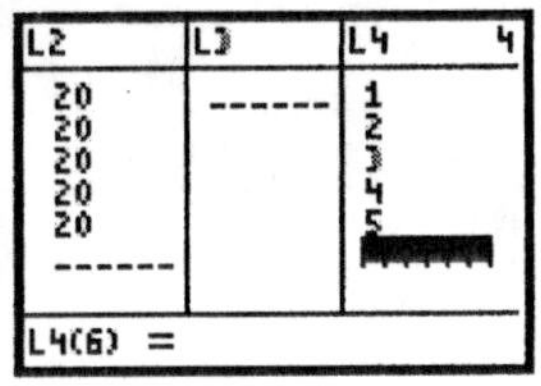	Enter the data into list L1. Enter 20's into L2. This is the expected value if consumers show no preference. Use the arrow keys to move from one list to the other. Enter 1, 2, 3, 4, 5 into L4 as the codes for Cherry, Strawberry, Orange, Lime and Grape. L3 will be used in our calculations.

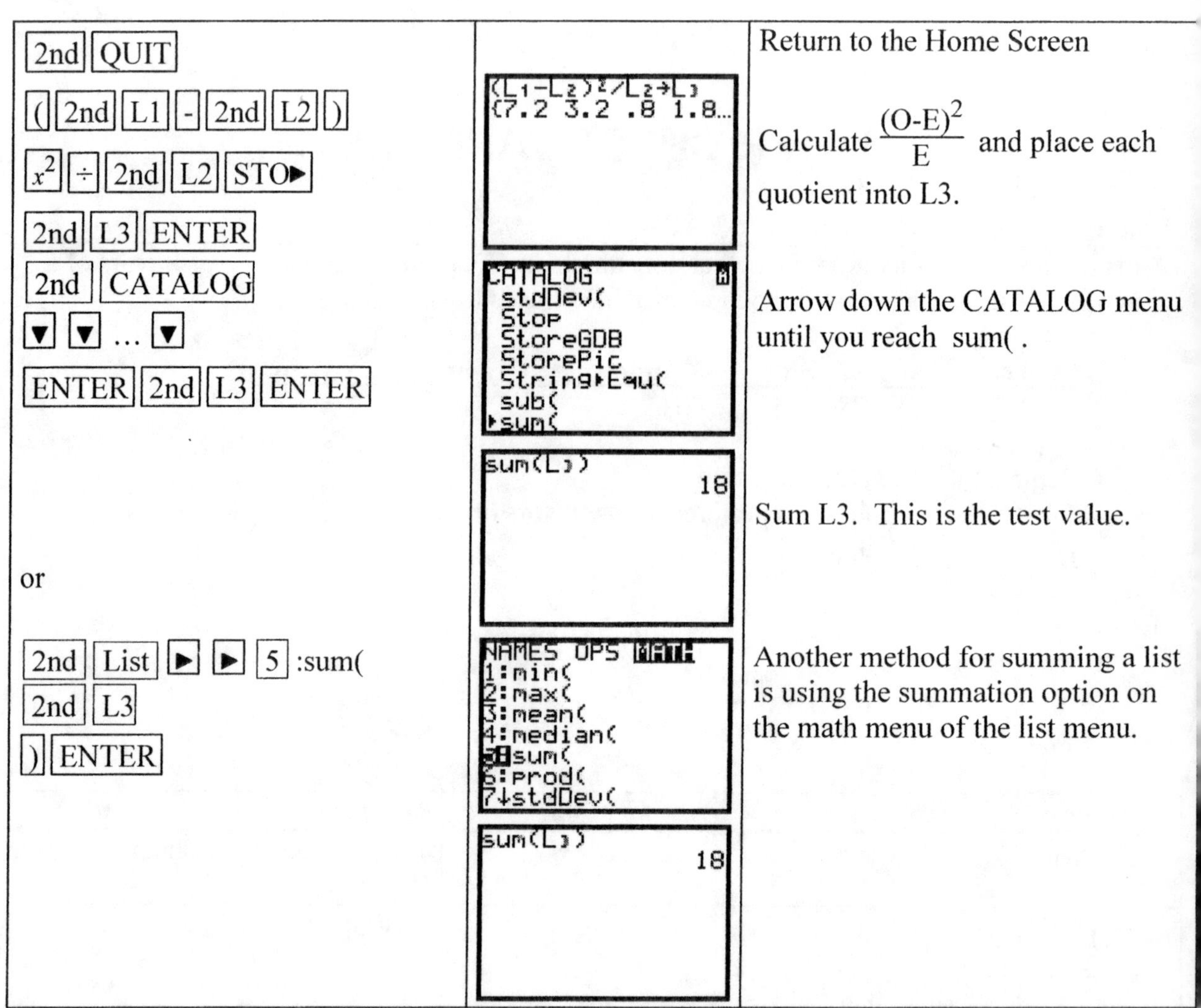

2nd QUIT		Return to the Home Screen
(2nd L1 - 2nd L2) x^2 ÷ 2nd L2 STO▶ 2nd L3 ENTER	(L1-L2)²/L2→L3 {7.2 3.2 .8 1.8…	Calculate $\frac{(O-E)^2}{E}$ and place each quotient into L3.
2nd CATALOG ▼ ▼ … ▼	CATALOG stdDev(Stop StoreGDB StorePic String▸Equ(sub(▸sum(	Arrow down the CATALOG menu until you reach sum(.
ENTER 2nd L3 ENTER	sum(L3) 18	Sum L3. This is the test value.
or		
2nd List ▶ ▶ 5 :sum(2nd L3) ENTER	NAMES OPS MATH 1:min(2:max(3:mean(4:median(5:sum(6:prod(7↓stdDev(	Another method for summing a list is using the summation option on the math menu of the list menu.
	sum(L3) 18	

What a person observes will never be exactly what one would expect So the questions is, are these differences significant (a preference exists), or are they due to chance? So we complete this test of hypothesis by comparing the test value of 18 to the chi-squared value found in the table for 4 degrees of freedom. The critical value is 9.488. Since the test value is greater than the critical value found from Table G, the decision is to reject the null hypothesis in favor of the alternative. Thee is enough evidence to reject the claim that consumers show no preference for the flavors in favor of the alternative hypothesis that there is a difference in preference for the difference flavors.

Solution (B):

Keystrokes	*Screen Display*	*Comments*
CLEAR Y= CLEAR 2nd DRAW 1 :ClrDraw ENTER		Turn the calculator on and clear the Home Screen. Delete all functions from the Y= list. Clear all drawings.

WINDOW 0 ENTER 6 ENTER 1 ENTER 0 ENTER 50 ENTER 5 2nd QUIT	WINDOW Xmin=0 Xmax=6 Xscl=1 Ymin=0 Ymax=50 Yscl=5 Xres=1	Enter the graph screen dimensions.
2nd STAT PLOT 1 :Plot1...Off ENTER	STAT PLOTS 1:Plot1...Off L4 L1 2:Plot2...Off L4 L2 3:Plot3...Off L1 L2 4↓PlotsOff	Select Plot1. Use the arrow keys to select appropriate options.
▼ ► ENTER ▼ 2nd L4 ENTER 2nd L1 ▼ ENTER	Plot1 Plot2 Plot3 On Off Type: Xlist:L4 Ylist:L1 Mark: ▫ + ·	Turn Plot1 on and select the type of graph and lists desired.
GRAPH		Graph.
2nd STAT PLOT 2 :Plot2...Off ENTER ▼ ► ENTER ▼ 2nd L4 ENTER 2nd L2 ▼ ENTER	Plot1 Plot2 Plot3 On Off Type: Xlist:L4 Ylist:L2 Mark: ▫ + ·	Turn Plot2 on and use the arrow keys to select the type of graph and lists desired.
GRAPH		Graph.

It can be seen from the graph that there are more than what was expected who prefer cherry and strawberry and less who prefer orange, lime or grape.

Example 2

(A) Test the hypothesis:

H_0 : The opinion about the procedure is independent of the profession.

H_1 : The opinion about the procedure is dependent on the profession.

Group	Prefer new procedure	Prefer old procedure	No preference
Nurses	100	80	20
Doctors	50	120	30

(B) Graph the Chi-square distribution with the *P*-Value shaded.

Solution:

Keystrokes	*Screen Display*	*Comments*
CLEAR		Turn the calculator on and clear the Home Screen.
STAT ► ► ALPHA C :χ^2-Test...	EDIT CALC TESTS 0↑2-SampTInt... A:1-PropZInt... B:2-PropZInt... C:χ²-Test... D:2-SampFTest... E:LinRegTTest... F:ANOVA(	Select the Chi-square test.
▼ ▼ :Calculate ENTER	χ²-Test Observed: [A] Expected: [B] Calculate Draw	The observed values are in matrix A. The expected values will be placed in matrix B.
▼ ▼ ► ENTER	χ²-Test χ²=26.66666667 p=1.6195968E-6 df=2 χ²=26.6667 p=0	The *P*-Value is 0.0000016. This value is less than the level of significance. Reject the null hypothesis. There is enough evidence to conclude that the procedure is dependent on the profession. Notice that no shading is observed since the area to be shaded is so small.

EXERCISE SET 11

1. A staff member of an emergency medical service wishes to determine whether the number of accidents is equally distributed during the week. A week was selected at random, and the following data were obtained.

Day	**Monday**	**Tuesday**	**Wednesday**	**Thursday**	**Friday**	**Saturday**	**Sunday**
Number of accidents	26	24	22	19	38	41	19

The hypotheses are:

H_0 : The number of accidents is equally distributed during the week.

H_1 : The number of accidents is not equally distributed during the week.

(A) Find the test value for this test.

(B) Graph the Chi-square distribution with the *P*-Value area shaded.

(C) Graph the observed and expected values for days of the week.

2. A bank manager wishes to see whether there is any preference in the times that customers use the bank. Six specific hours are selected, and the number of customers visiting the bank during each hour are shown below.

(A) At $\alpha = 0.01$, do the customers show a preference for specific times?
(B) Graph the Chi-square distribution with the *P*-Value area shaded.

Time	**10:00**	**11:00**	**12:00**	**1:00**	**2:00**	**3:00**	**4:00**	**5:00**	**6:00**	**7:00**
Number of customers	32	33	42	36	28	19	25	43	39	18

The hypotheses are:

H_0 : The number of customers is equally distributed throughout the day.

H_1 : The number of customers is not equally distributed throughout the day.

(A) Find the test value for this test.

(B) Graph the Chi-square distribution with the P-Value area shaded.

(C) Graph the observed and expected values for times.

3. A researcher wishes to see whether the age of an individual is related to coffee consumption. A sample of 152 people is selected, and they are classified as shown in the table below.

	Coffee consumption		
Age	**Low**	**Moderate**	**High**
21-30	18	16	12
31-40	9	15	27
41-50	8	12	10
51 and over	13	10	9

(A) At $\alpha = 0.05$, is there a relationship between coffee consumption and age?

(B) Graph the Chi-square distribution with the *P*-Value area shaded.

4. A study is being conducted to determine whether there is a relationship between jogging and blood pressure. A random sample of 210 subjects is selected and classified as shown below.

	Coffee consumption		
Age	**Low**	**Moderate**	**High**
21-30	18	16	12
31-40	9	15	27
41-50	5	12	10
51 and over	13	9	6

(A) At $\alpha = 0.05$, is there a relationship between jogging and blood pressure?

(B) Graph the Chi-square distribution with the *P*-Value area shaded.

CHAPTER 12

ANALYSIS OF VARIANCE

Example 1

A researcher wishes to try three different techniques to lower the blood pressure of individuals diagnosed with high blood pressure. The subjects are randomly assigned to three groups: The first group takes medication, the second group exercises, and the third group follows a special diet. After four weeks, the reduction in each person's blood pressure is recorded. At the 0.05 level of significance, test the claim that there is no difference among the means.

Medication	Exercise	Diet
10	6	5
12	8	9
9	3	12
15	0	8
13	2	4

Solution (A):

Keystrokes	*Screen Display*	*Comments*
[CLEAR]		Turn the calculator on and clear the Home Screen.
[STAT] [4] :ClrList [2nd] [L1] [,] [2nd] [L2] [,] [2nd] [L3] [ENTER] [STAT] [1] :Edit [10] [ENTER] [12] [ENTER] etc. [▶] [6] [ENTER] [8] [ENTER] etc. [▶] [5] [ENTER] [9] [ENTER] etc.	EDIT CALC TESTS 1:Edit… 2:SortA(3:SortD(4:ClrList 5:SetUpEditor L1 L2 L3 10 6 5 12 8 9 9 3 12 15 0 8 13 2 4 L3(6) =	Clear the lists and enter the data.

STAT ▶ ▶ ALPHA F :ANOVA(2nd L1 , 2nd L2 , 2nd L3)	ANOVA(L1,L2,L3)	Select ANOVA from the statistics hypothesis tests menu. The data was stored in L1, L2, and L3.

Example 2

A researcher wishes to try three different techniques to lower the blood pressure of individuals diagnosed with high blood pressure. The subjects are randomly assigned to three groups: The first group takes medication, the second group exercises, and the third group follows a special diet. After four weeks, the reduction in each person's blood pressure is recorded. At the 0.05 level of significance, test the claim that there is no difference among the means.

Medication	Exercise	Diet
10	6	5
12	8	9
9	3	12
15	0	8
13	2	4

Solution (A):

Keystrokes	*Screen Display*	*Comments*
CLEAR		Turn the calculator on and clear the Home Screen.
STAT 4 :ClrList 2nd L1 , 2nd L2 , 2nd L3 ENTER STAT 1 :Edit 10 ENTER 12 ENTER etc. ▶ 6 ENTER 8 ENTER etc. ▶ 5 ENTER 9 ENTER etc.	EDIT CALC TESTS 1:Edit… 2:SortA(3:SortD(4:ClrList 5:SetUpEditor L1: 10, 12, 9, 15, 13 L2: 6, 8, 3, 0, 2 L3: 5, 9, 12, 8, 4 L3(6) =	Clear the lists and enter the data.

STAT ▶ ▶ ALPHA F :ANOVA(2nd L1 , 2nd L2 , 2nd L3)	ANOVA(L1,L2,L3)	Select ANOVA from the statistics hypothesis tests menu. The data was stored in L1, L2, and L3.
	One-way ANOVA F=9.167938931 p=.0038313169 Factor df=2 SS=160.133333 ↓ MS=80.0666667 One-way ANOVA ↑ MS=80.0666667 Error df=12 SS=104.8 MS=8.73333333 Sxp=2.95522137	Use the arrow keys to see the rest of the output. The *P*-Value is 0.004. This is less than the level of significance. Reject the null hypothesis of equal means. There is enough evidence to conclude that at least one mean is different from the others.
ENTER	One-way ANOVA F=9.167938931 p=.0038313169 Factor df=2 SS=160.133333 ↓ MS=80.0666667 One-way ANOVA ↑ MS=80.0666667 Error df=12 SS=104.8 MS=8.73333333 Sxp=2.95522137	Use the arrow keys to see the rest of the output. The *P*-Value is 0.004. This is less than the level of significance. Reject the null hypothesis of equal means. There is enough evidence to conclude that at least one mean is different from the others.

Example 3

A researcher wishes to see whether the type of gasoline used and the type of automobile driven have any effect on gasoline consumption. Two types of gasoline, regular and high-octane, will be used, and two types of automobiles, two-wheel- and four-wheel-drive, will be used in each group. There will be two automobiles in each group, for a total of eight automobiles used. Draw a graph plotting the means of each group, analyzing the graph, and interpreting the results. The means for each cell are shown in the chart below.

	Type of Automobile	
Gas	**Two-wheel-drive**	**Four-wheel-drive**
Regular	$\overline{X} = \frac{26.7 + 25.2}{2} = 25.95$	$\overline{X} = \frac{28.6 + 29.3}{2} = 28.95$
High-octane	$\overline{X} = \frac{32.3 + 32.8}{2} = 32.55$	$\overline{X} = \frac{26.1 + 24.2}{2} = 25.15$

Keystrokes	*Screen Display*	*Comments*
[CLEAR]		Turn the calculator on and clear the Home Screen.
[2nd] [DRAW] [1] :ClrDraw [ENTER] [Y=] [CLEAR] [ENTER] [CLEAR] [ENTER] etc. [STAT] [4] :ClrList [2nd] [L1] [,] [2nd] [L2] [,] [2nd] [L3] [ENTER] [STAT] [1] :Edit [1] [ENTER] [3] [ENTER] [▶] [29.95] [ENTER] [28.95] [ENTER] [▶] [32.55] [ENTER] [25.15] [ENTER]	EDIT CALC TESTS 1:Edit… 2:SortA(3:SortD(4:ClrList 5:SetUpEditor	Clear all drawings and all functions from the Y= list. Get the statistics menu and select 1: Edit… Clear the lists and enter the data. We will enter the numbers 1 and 3 in the first column so that when we set the WINDOW at Xmin=0 and Xmax=4 the graph will appear in the center of the screen. Set the WINDOW at: Xmin=0, Xmax=4, Xscl=1, Ymin=24, Ymax=34, Yscl=1, Xres=1.
[WINDOW] [0] [ENTER] [4] [ENTER] [1] [ENTER] [25] [ENTER] [35] [ENTER] [1] [ENTER] [1] [ENTER] [2nd] [QUIT]	WINDOW Xmin=0 Xmax=4 Xscl=1 Ymin=25 Ymax=35 Yscl=1 Xres=1	Set the Window dimensions. Return to the Home Screen.

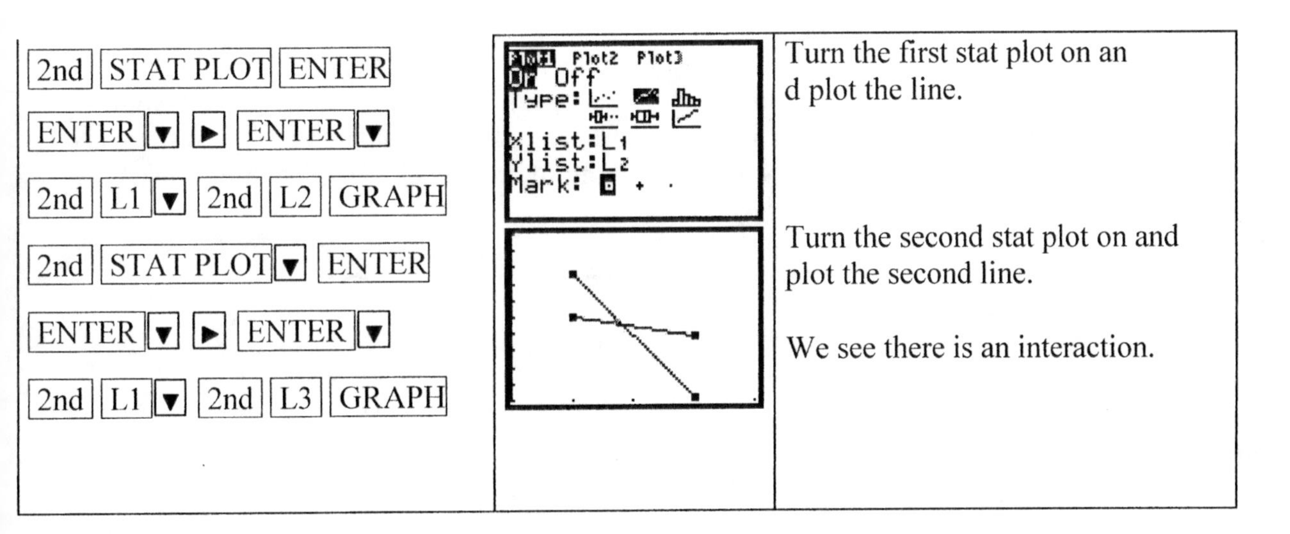

EXERCISE SET 12

1. A plant owner wants to see whether the average time (in minutes) it takes his employees to commute to work is different for three groups. The data are shown here. At the 0.05 level of significance, can the owner conclude that there is a significant difference among the means?

Managers	Salespeople	Stock clerks
35	9	15
18	3	6
27	12	27
24	6	42
	14	18
	8	
	21	

2. Four hospitals are being compared to see whether there is any significant difference in the mean number of operations performed in each. A sample of six days provided the following number of operations performed each day. Test the claim at the 0.01 level of significance that there is no difference in the means.

Hospital A	Hospital B	Hospital C	Hospital D
8	4	5	10
5	9	6	12
6	3	3	13
3	1	7	
	0	8	
	1		

3. A medical researcher wishes to test the effects of two diets and the time of day on the sodium level in a person's blood. Eight people are randomly selected, and two are randomly assigned to each of the four groups. Analyze the graph to see if there is an interaction between the time of day and the sodium level. The data are shown here.

	Diet Type	
Time	**I**	**II**
8:00 A.M.	135 145	138 141
8:00 P.M.	155 162	171 191

4. A teacher wishes to test the math anxiety level of her students in two classes at the beginning of the semester. The classes are Calculus I and Statistics. Furthermore, she wishes to see whether here is a difference owing to the students' ages. Math anxiety is measured by the score on a 100-point anxiety test. Analyze the graph to see if there is an interaction between the subject content and the age of the students. The data are shown here.

	Class	
Age	**Calculus I**	**Statistics**
Under 20	43, 52, 61, 57, 55	19, 20, 31, 36, 24
20 or over	56, 55, 42, 48, 61	63, 78, 67, 71, 75

CHAPTER 13

NONPARAMETRIC STATISTICS

Example 1

A convenience store owner hypothesizes that the median number of snow cones he sells per day is 40. A random sample of 20 days yields the following data for the number of snow cones sold each day.

18	43	40	16	22	30	29	32	37	36
39	34	39	45	28	36	40	34	39	52

Find the test value using the sign test for the hypotheses:

H_0: median = 40 (claim) H_1: median $\neq$ 40

Solution:

Keystrokes	*Screen Display*	*Comments*
[CLEAR]		Turn the calculator on and clear the Home Screen.
[STAT] [4] :ClrList [2nd] [L1] [,] [2nd] [L2] [,] [2nd] [L3] [,] [2nd] [L4]	EDIT CALC TESTS 1:Edit… 2:SortA(3:SortD(4:ClrList 5:SetUpEditor	Clear the lists.
[ENTER] [STAT] [1] :Edit [18] [ENTER] [43] [ENTER] etc.	L1 L2 L3 1 28 36 40 34 39 52 L1(21) =	Enter the data.
[STAT] [2] :SortA([2nd] [L1] [)]	EDIT CALC TESTS 1:Edit… 2:SortA(3:SortD(4:ClrList 5:SetUpEditor	Rank the data ascending using SortA(from the STAT EDIT menu.
[ENTER]	ClrList L1 Done SortA(L1) Done	

Keystrokes	Screen	Explanation
[STAT] [1] :Edit	L1 L2 L3 1 36 36 37 39 39 39 40 L1(16) =40	The first value of the median occurs in position 16. Hence there are 15 pieces of data less than the median.
[STAT] [3] :SortD([2nd] [L1] [)] [ENTER]	ClrList L1 Done SortA(L1) Done SortD(L1) Done	Sort the list in descending order.
[STAT] [1] :Edit	L1 L2 L3 1 52 45 43 40 40 39 39 L1(4)=40	The first value of the median in the descending list occurs in position 4. Hence there are 3 pieces of data greater than the median. The test value is 3 since it is the smaller value.

Example 2

A medical researcher believed the number of ear infections in swimmers can be reduced if the swimmers use earplugs. A sample of 10 people was selected, and the number of infections for a four-month period was recorded. During the first two months, the swimmers did not use the earplugs; during the second two months, they did. At the beginning of the second two-month period, each swimmer was examined to make sure that no infections were present. The data are shown below.

(A) Find the test value for the paired-sample sign test.

(B) Find the test value for the Wilcoxon signed-rank test.

Number of ear infections

Swimmer	Before	After
A	3	2
B	0	1
C	5	4
D	4	0
E	2	1
F	4	3
G	3	1
H	5	3
I	2	2
J	1	3

Solution (A):

Keystrokes	*Screen Display*	*Comments*
CLEAR		Turn the calculator on and clear the Home Screen.
STAT 4 :ClrList 2nd L1 , 2nd L2 , 2nd L3 ENTER	EDIT CALC TESTS 1:Edit... 2:SortA(3:SortD(4:ClrList 5:SetUpEditor	Clear the lists.
STAT 1 :Edit 3 ENTER 0 ENTER etc. ▶ 2 ENTER 1 ENTER etc.	L1 L2 L3 2 L2(11) =	Enter data into L1 and L2.
2nd QUIT		Return to the Home Screen.
2nd L1 - 2nd L2 STO▶ 2nd L3 ENTER	L1-L2→L3 {1 -1 1 4 1 1 2...	Find the difference by subtracting L2 from L1 and storing in L3.
STAT 2 :SortA(2nd L3 , 2nd L1 , 2nd L2) ENTER	L1-L2→L3 {1 -1 1 4 1 1 2... SortA(L3,L1,L2) Done	Sort the lists ascending using L3 as the first list. L1 and L2 will also be sorted keeping the pairs together.
STAT 1 :Edit ▶ ▶ ▼ ... ▼	L1 L2 L3 3 L3(3) =0	Examine the list to see that the first 0 occurs in position 3. Hence there are 2 negative signs.
STAT 3 :SortD(2nd L3 , 2nd L1 , 2nd L2) ENTER	L1-L2→L3 {1 -1 1 4 1 1 2... SortA(L3,L1,L2) Done SortD(L3,L1,L2) Done	Sort the lists descending using L3 as the first list. L1 and L2 will also be sorted keeping the pairs together.
STAT 1 :Edit ▶ ▶ ▼ ... ▼	L1 L2 L3 3 L3(8) =0	Examine the list to see that the first 0 occurs in position 8. Hence there are 7 positive signs. The smaller of the number of negative and positive signs is 2. This is the test value.

Solution (B):

Keystrokes	*Screen Display*	*Comments*
[CLEAR]		Turn the calculator on and clear the Home Screen.
[STAT] [4] :ClrList [2nd] [L1] [,] [2nd] [L2] [,] [2nd] [L3] [,] [2nd] [L4] [,] [2nd] [L5] [ENTER] [STAT] [1] :Edit [3] [ENTER] [0] [ENTER] etc. [▶] [2] [ENTER] [1] [ENTER] etc. [2nd] [QUIT] [CLEAR]	EDIT CALC TESTS 1:Edit… 2:SortA(3:SortD(4:ClrList 5:SetUpEditor L1 L2 L3 2 L2(11) =	Clear the lists. Enter data into L1 and L2 if it is not already entered. Return to the Home Screen.
[2nd] [L1] [-] [2nd] [L2] [STO▶] [2nd] [L3] [ENTER]	L1−L2→L3 {1 -1 1 4 1 1 2…	Find the difference by subtracting L2 from L1 and storing in L3, if this has not already been done.
MATH [▶] [1] :abs([2nd] [L3] [)] [STO▶] [2nd] [L4] [ENTER]	L1−L2→L3 {1 -1 1 4 1 1 2… abs(L3)→L4 {1 1 1 4 1 1 2 …	Find the absolute value of L3 and store in L4. If you are continuing this part from Part (A), then the numbers displayed will be the sorted as the sorded descending list of L3.
[STAT] [2] :SortA([2nd] [L4] [,] [2nd] [L1] [,] [2nd] [L2] [,] [2nd] [L3] [)] [ENTER]	L1−L2→L3 {1 -1 1 4 1 1 2… abs(L3)→L4 {1 1 1 4 1 1 2 … SortA(L4,L1,L2,L3) Done	Sort the lists ascending using L4 as the first list. L1, L2 and L3 will also be sorted keeping the pairs together.

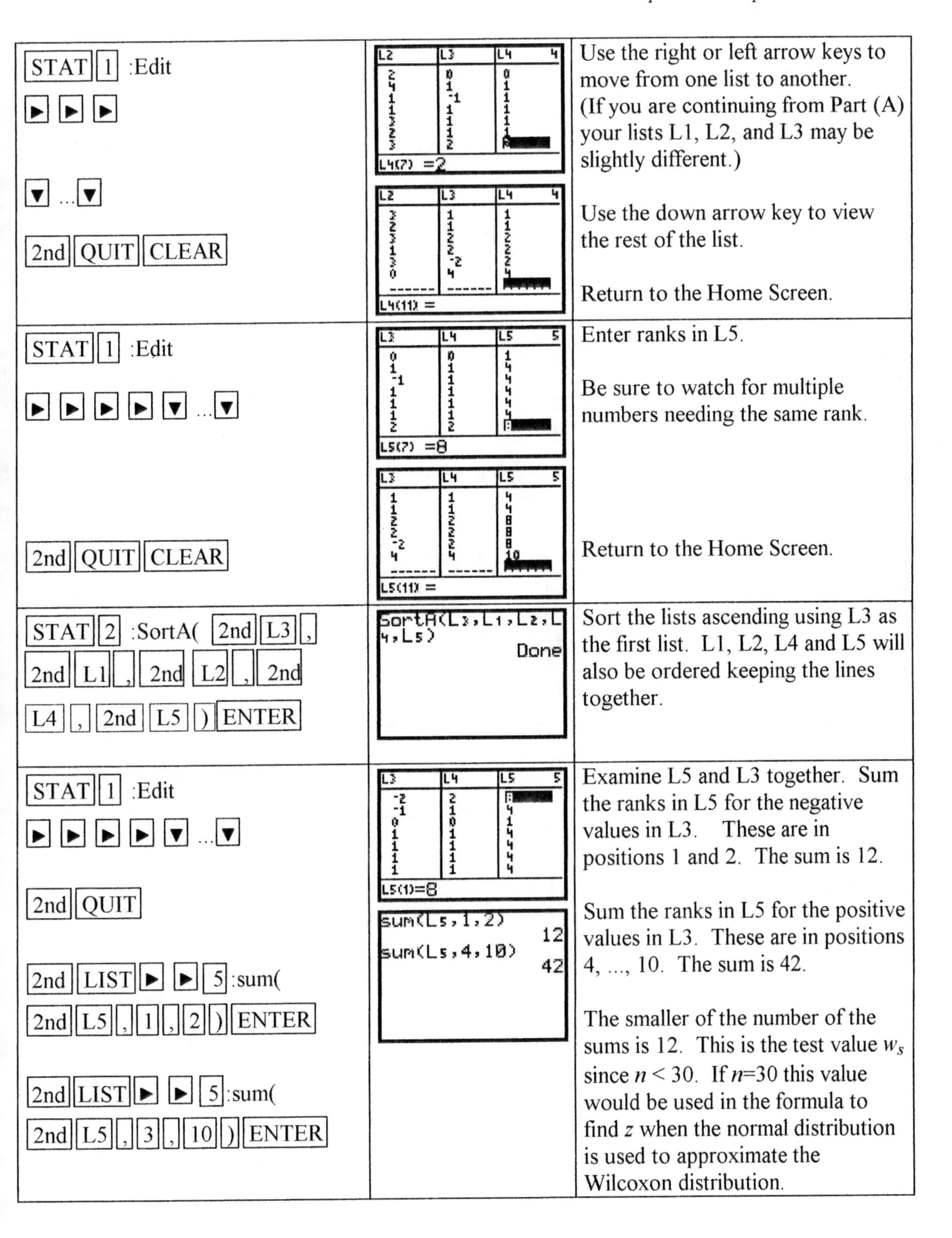

Keystrokes	Explanation
STAT 1 :Edit ▶ ▶ ▶ ▼ ... ▼ 2nd QUIT CLEAR	Use the right or left arrow keys to move from one list to another. (If you are continuing from Part (A) your lists L1, L2, and L3 may be slightly different.) Use the down arrow key to view the rest of the list. Return to the Home Screen.
STAT 1 :Edit ▶ ▶ ▶ ▶ ▼ ... ▼ 2nd QUIT CLEAR	Enter ranks in L5. Be sure to watch for multiple numbers needing the same rank. Return to the Home Screen.
STAT 2 :SortA(2nd L3 , 2nd L1 , 2nd L2 , 2nd L4 , 2nd L5) ENTER	Sort the lists ascending using L3 as the first list. L1, L2, L4 and L5 will also be ordered keeping the lines together.
STAT 1 :Edit ▶ ▶ ▶ ▶ ▼ ... ▼ 2nd QUIT 2nd LIST ▶ ▶ 5 :sum(2nd L5 , 1 , 2) ENTER 2nd LIST ▶ ▶ 5 :sum(2nd L5 , 3 , 10) ENTER	Examine L5 and L3 together. Sum the ranks in L5 for the negative values in L3. These are in positions 1 and 2. The sum is 12. Sum the ranks in L5 for the positive values in L3. These are in positions 4, ..., 10. The sum is 42. The smaller of the number of the sums is 12. This is the test value w_s since $n < 30$. If n=30 this value would be used in the formula to find z when the normal distribution is used to approximate the Wilcoxon distribution.

Example 3

Two independent samples of army and marine recruits are selected, and the time in minutes it takes each recruit to complete an obstacle course is recorded as shown in the table. Find n_1 and n_2 for testing if there is a difference in the times it takes the recruits to complete the course using the Wilcoxon rank sum test.

Army	15	18	16	17	13	22	24	17	19	21	26	28
Marines	14	9	16	19	10	12	11	8	15	18	25	

Solution:

Keystrokes	*Screen Display*	*Comments*
[CLEAR] or [2nd][QUIT][CLEAR]		Turn the calculator on and clear the Home Screen.
[STAT][4] :ClrList [2nd][L1][,] [2nd][L2][,][2nd][L3] [ENTER]	EDIT CALC TESTS 1:Edit... 2:SortA(3:SortD(4:ClrList 5:SetUpEditor	Clear the lists.
[STAT][1] :Edit [15][ENTER][18][ENTER] etc. [►] [1][ENTER] [1][ENTER] etc. [2nd][QUIT]	L1 L2 L3 2 12 2 11 2 8 2 15 2 18 2 25 2 L2(24)=	Enter all data into L1 with a corresponding number in L2 as the codes: 1=Army 2=Marine Return to the Home Screen.
[STAT][2] :SortA([2nd][L1][,] [2nd][L2][)] [ENTER]	SortA(L1,L2) Done	Sort the data in L1 with corresponding codes in L2.
[STAT][1] :Edit [►] [►] [1][ENTER][2][ENTER] etc. [2nd][QUIT]	L1 L2 L3 3 18 2 14.5 18 1 14.5 19 2 16.5 19 1 16.5 21 1 22 1 19 24 1 20 L3(18)=18	Enter ranks in L3. Be sure to use halves where appropriate. Return to the Home Screen.

STAT 2 :SortA(2nd L2 , 2nd L1 , 2nd L3) ENTER	SortA(L2,L1,L3) Done	Sort ascending by groups.
STAT 1 :Edit ▶ ▼ ... ▼	L1 L2 L3 2 28 1 23 18 2 14.5 16 2 10.5 15 2 8.5 19 2 16.5 14 2 7 12 2 5 L2(13) =2	Use the down arrow key to see where group 2 begins. Group 1 (Army) are items 1 through 12 Group 2 (Marines) are items 13 through 23).
2nd QUIT 2nd LIST ▶ ▶ 5 :sum(L3 , 1 , 12) ENTER 2nd QUIT 2nd LIST ▶ ▶ 5 :sum(L3 , 13 , 23) ENTER	sum(L3,1,12) 183 sum(L3,13,23) 93	Return to the Home Screen. Sum L3 from item 1 through 12 and sum L3 from item 13 through 23 using sum(from the list math menu. The smaller sum is 93. The number of data values is 13. So, $n_1 = 11$ and $n_2 = 12$ in the formula.

Example 4

A researcher tests three different brands of breakfast drinks to see how many milliequivalents of potassium per quart each contains. The following data are obtained.

Brand A	Brand B	Brand C
4.7	5.3	6.3
3.2	6.4	8.2
5.1	7.3	6.2
5.2	6.8	7.1
5.0	7.2	6.6

Find the test value for the Kruskal-Wallis test.

Solution:

Keystrokes	*Screen Display*	*Comments*
CLEAR		Turn the calculator on and clear the Home Screen.

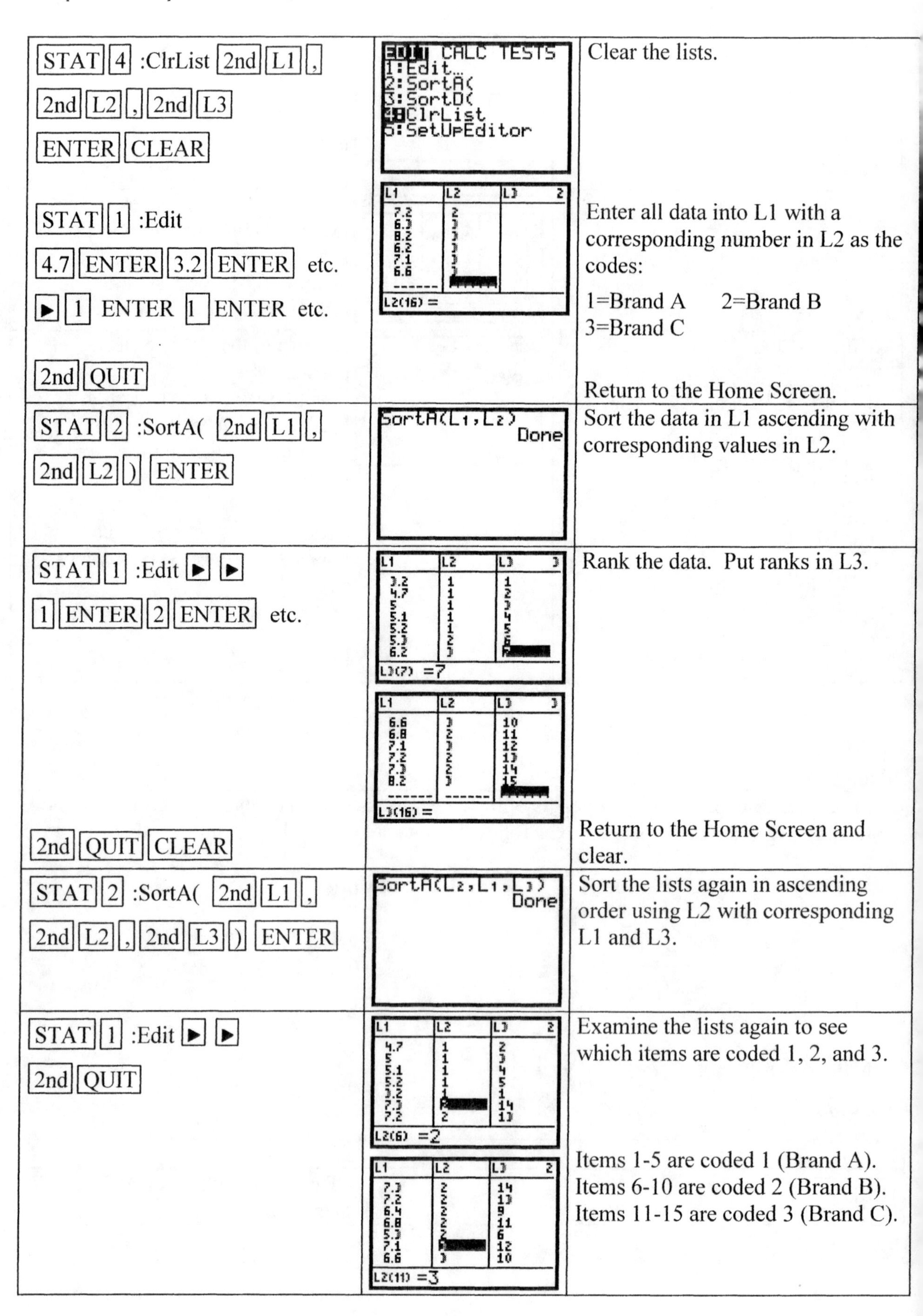

Keystrokes	Screen	Explanation
STAT 4 :ClrList 2nd L1 , 2nd L2 , 2nd L3 ENTER CLEAR	EDIT CALC TESTS 1:Edit… 2:SortA(3:SortD(4:ClrList 5:SetUpEditor	Clear the lists.
STAT 1 :Edit 4.7 ENTER 3.2 ENTER etc. ▶ 1 ENTER 1 ENTER etc. 2nd QUIT		Enter all data into L1 with a corresponding number in L2 as the codes: 1=Brand A 2=Brand B 3=Brand C Return to the Home Screen.
STAT 2 :SortA(2nd L1 , 2nd L2) ENTER	SortA(L1,L2) Done	Sort the data in L1 ascending with corresponding values in L2.
STAT 1 :Edit ▶ ▶ 1 ENTER 2 ENTER etc. 2nd QUIT CLEAR		Rank the data. Put ranks in L3. Return to the Home Screen and clear.
STAT 2 :SortA(2nd L1 , 2nd L2 , 2nd L3) ENTER	SortA(L2,L1,L3) Done	Sort the lists again in ascending order using L2 with corresponding L1 and L3.
STAT 1 :Edit ▶ ▶ 2nd QUIT		Examine the lists again to see which items are coded 1, 2, and 3. Items 1-5 are coded 1 (Brand A). Items 6-10 are coded 2 (Brand B). Items 11-15 are coded 3 (Brand C).

[2nd] [LIST] [▶] [▶] [5]:sum([2nd] [L3] [,] [1] [,] [5] [)] [ENTER] [2nd] [LIST] [▶] [▶] [5]:sum([2nd] [L3] [,] [6] [,] [10] [)] [ENTER] [2nd] [LIST] [▶] [▶] [5]:sum([2nd] [L3] [,] [11] [,] [15] [)] [ENTER]	sum(L3,1,5) 15 sum(L3,6,10) 53 sum(L3,11,15) 52	Sum the ranks for the items separately for each category. N=15, R_1=15, R_2=53, R_3=52 n_1= 5, n_2= 5, n_3= 5
[CLEAR] [12] [÷] [(] [15] [x] [(] [15] [+] [1] [)] [)] [x] [(] [15] [x^2] [÷] [5] [+] [53] [x^2] [÷] [5] [+] [52] [x^2] [÷] [5] [)] [-] [3] [(] [15] [+] [1] [)] [ENTER]	12/(15*(15+1))*(15²/5+53²/5+52²/5)-3(15+1) 9.38	Clear the Home Screen. Calculate the test value. The test value is 9.38.

Example 5

Two students were asked to rate eight different textbooks for a specific course on an ascending scale from 0 to 20 points. Points were assigned for each of several categories, such as reading level, use of illustrations, and use of color. Find the Spearman rank correlation coefficient. Use three decimal places. The following data are obtained.

Textbook	Student 1's rating	Student 2's rating
A	4	4
B	10	6
C	18	20
D	20	14
E	12	16
F	2	8
G	5	11
H	9	7

Solution:

Keystrokes	*Screen Display*	*Comments*
[CLEAR]		Turn the calculator on and clear the Home Screen.

Keystrokes	Screen	Explanation
STAT 4 :ClrList 2nd L1 , 2nd L2 , 2nd L3 , 2nd L4 , 2nd L5 ENTER STAT 1 :Edit ▶ ... ▶ 4 ENTER 10 ENTER etc. ▶ ▶ 4 ENTER 6 ENTER etc. 2nd QUIT CLEAR	EDIT CALC TESTS 1:Edit... 2:SortA(3:SortD(4:ClrList 5:SetUpEditor L1: 4 10 18 20 12 2 5 L2: L3: 4 6 20 14 16 8 11 L2(1)=	Clear the lists. Move to list L1. Enter data into L1 and L3. L2 and L4 will be used for ranks. Return to the Home Screen and clear.
STAT 2 :SortA(2nd L1 , 2nd L3) ENTER 2nd QUIT CLEAR	L1: 2 4 5 9 10 12 18 L2: 1 2 3 4 5 6 7 L3: 8 4 11 7 6 16 20 L2(1)=1 L1: 5 9 10 12 18 20 L2: 3 4 5 6 7 8 L3: 11 7 6 16 20 14 L2(9) =	Sort L1 in ascending order keeping the correspondence with L3. Rank L1. Put the ranks in L2. Return to the Home Screen.
STAT 2 :SortA(2nd L3 , 2nd L1 , 2nd L2) ENTER 2nd QUIT	L2: 4 1 3 8 6 7 L3: 7 8 11 14 16 20 L4: 3 4 5 6 7 8 L4(9) =	Sort L3 using the keeping the correspondences with L1 and L2. Rank L3. Put the ranks in L4. Return to the Home Screen.
(2nd L2 - 2nd L4) x^2 STO▲ 2nd L5) ENTER	(L2-L4)²→L5 {1 9 1 9 4 4 1 ...	Subtract the ranks, square them, and place in L5.
2nd LIST ▶ ▶ 5 :sum(2nd L5) ENTER	(L2-L4)²→L5 {1 9 1 9 4 4 1 ... sum(L5) 30	Sum L5. The sum is 30. Σd^2 = 30.
1 - 6 x 30 ÷ (8 (8 x^2 - 1)) ENTER	(L2-L4)²→L5 {1 9 1 9 4 4 1 ... sum(L5) 30 1-6*30/(8(8²-1)) .6428571429	Substitute in the formula and calculate to find $r_s = 0.643$.

EXERCISE SET 13

1. A real estate agent suggest that the median rent for a one-bedroom apartment in Blue View is $325 per month. A sample of one-bedroom apartments shows the following monthly rent for a one-bedroom apartment. Find the test value for the sign test for the hypotheses:

   H_0: median = $325 (claim) versus H_1: median ≠? $325

$420	$460	$514	$405
320	435	531	450
560	309	312	350

2. A government economist estimates that the median cost per pound of beef is $5.00. A sample of 22 livestock buyers shows the following costs per pound of beef. Find the test value for testing the hypothesis that the median is $5.00.

$5.35	5.16	4.97	4.83	5.05	5.19
4.78	4.93	4.86	5.00	4.63	5.06
5.19	5.00	5.05	5.10	5.16	5.25
5.16	5.42	5.13	5.27		

3. A study was conducted to see whether a certain diet medication had an effect on the weights (in pounds) of eight women. Their weights were taken before and six weeks after daily administration of the medication. The data are shown here. Find the test value for testing that the medication had an effect (increase or decrease) on the weights of the women.

Subject	**A**	**B**	**C**	**D**	**E**	**F**	**G**	**H**
Weight Before	187	163	201	158	139	143	198	154
Weight After	178	162	188	156	133	150	175	150

   (A) Find the test value for the paired-sample sign test for testing that the medication had an effect (increase or decrease) on the weights of the women.

   (B) Find the test value for the Wilcoxon signed-rank test

4. Two different laboratory machines measure the sodium content (in milligrams) of the same 10 blood samples. The data are shown below.

Sample	**1**	**2**	**3**	**4**	**5**	**6**	**7**	**8**	**9**	**10**
Machine 1	138	136	142	151	154	141	140	138	132	136
Machine 2	140	136	141	140	153	144	143	136	131	138

(A) Find the test value for testing the claim that both machines gave the same reading for the paired-sample sign test.
(B) Find the test value for the Wilcoxon signed-rank test

5. In a large city the number of crimes per week in five precincts is recorded for five weeks. The data are shown here. Find the test value for the Kruskal-Wallis test.

Precinct 1	Precinct 2	Precinct 3	Precinct 4	Precinct 5
105	87	74	56	103
108	86	83	43	98
99	91	78	52	94
97	93	74	58	89
92	82	60	62	88

6. Three brands of microwave dinners were advertised as low in sodium. Samples of the three different brands show the following milligrams of sodium. Find the test value for the Kruskal-Wallis test.

Brand A	Brand B	Brand C
810	917	893
702	912	790
853	952	603
703	958	744
892	893	623
732		743
713		609
613		

7. Eight music videos were ranked by teenagers and their parents on style and clarity, with 1 being the highest ranking. The data are shown below. Find the Spearman Rank Correlation Coefficient.

Music videos	A	B	C	D	E	F	G	H
Teenagers	4	6	2	8	1	7	3	5
Parents	1	7	5	4	3	8	2	6

8. The sociology department is selecting new textbooks for the next semester. Five instructors and three teaching assistants reviewed seven books and assigned each book rating points on a scale of 1 to 12, with 1 being the poorest and 12 being the best. Each book was rated on content, readability, etc. The data are shown below. Find the Spearman Rank Correlation Coefficient.

Textbooks	A	B	C	D	E	F	G
Instructors	5	8	12	3	11	9	1
Assistants	5	7	10	4	11	8	2

CHAPTER 14

SAMPLING

Example 1

Generate six random integers at least 0 and less than 30.

(A) Use a seed of 3.
(B) Use a seed of 5.

Solution (A):

METHOD 1

The random numbers produced by the calculator are always between 0 and 1. We want to set it so it will generate integers at least 0 and less than 30. To do this multiply the random number by 30 and take the integer of the result.

Keystrokes	*Screen Display*	*Comments*
CLEAR		Turn the calculator on and clear the Home Screen.
3 STO▶ MATH ▶ ▶ ▶ 1 :rand ENTER	3→rand 3	Store 3 as the seed. To do this get rand from the probability menu on the MATH menu.
MATH ▶ 3 :iPart(30 x MATH ▶ ▶ ▶ 1 :rand) ENTER ENTER ENTER ENTER ENTER ENTER	iPart(30*rand) 7 17 20 12 1 7	Get the integer part of the random number that has been multiplied by 30. The random numbers are 7, 17, 20, 12, 1, and 7.

METHOD 2

This method uses the randInt(function of the calculator. This function finds a specified number of integers between and including two specified integers.

Keystrokes	*Screen Display*	*Comments*
CLEAR		Turn the calculator on and clear the Home Screen.
3 STO▶ MATH ▶ ▶ ▶ 1 :rand ENTER	3→rand 3	Store 3 as the seed. To do this get rand from the probability menu on the MATH menu.

MATH ▶ ▶ ▶ 5 :randInt(0 , 29 , 6) ENTER	3→rand 3 randInt(0,29,6) {7 17 20 12 1 7}	Get the random integer function from the probability menu on the MATH menu. The lower limit is 0. The upper limit is 29. The number of random integers we wish is 6. Use the right arrow key to see the rest of the list if there are ... at the right side of the screen.

The random numbers are 7, 17, 20, 12, 1, and 7.

If you press ENTER again you will get the next six random numbers.

Solution (B):

METHOD 1

Keystrokes	*Screen Display*	*Comments*
CLEAR		Turn the calculator on and clear the Home Screen.
5 STO▶ MATH ▶ ▶ ▶ 1 :rand ENTER	5→rand 5	Store 5 as the seed. To do this get rand from the probability menu on the MATH menu.
MATH ▶ 3 :iPart(30 x MATH ▶ ▶ ▶ 1 :rand) ENTER ENTER ENTER ENTER ENTER ENTER	iPart(30*rand) 21 8 3 10 13 12	Get the integer part of the random number that has been multiplied by 30. The random numbers are 21, 8, 3, 10, 13, and 12.

METHOD 2

This method uses the randInt(function of the calculator. This function finds a specified number of integers between and including two specified integers.

Keystrokes	*Screen Display*	*Comments*
CLEAR		Turn the calculator on and clear the Home Screen.
5 STO▶ MATH ▶ ▶ ▶ 1 :rand ENTER	5→rand 5	Store 5 as the seed.

MATH ▶ ▶ ▶ 5 :randInt(0 , 29 , 6) ENTER	5→rand 5 randInt(0,29,6) {21 8 3 10 13 1...	The lower limit is 0. The upper limit is 29. The number of random integers we wish is 6. Use the right arrow key to see the rest of the list if there are ... at the right side of the screen. The random numbers are 21, 8, 3, 10, 13, and 12.

If you press ENTER again you will get the next six random numbers.

Example 2

Use the calculator's random number generator to simulate 60 tosses of a six-sided die having faces labeled with the integers 1 to 6. Use 3 as the seed.

Solution (A):

The random numbers produced by the calculator are always between 0 and 1. We want to set it so it will generate integers 1, 2, 3, 4, 5 and 6. To do this multiply the random number by 6, take the integer of the result and add 1.

METHOD 1

Keystrokes	*Screen Display*	*Comments*
CLEAR		Turn the calculator on and clear the Home Screen.
3 STO▶ MATH ▶ ▶ ▶ 1 :rand ENTER	3→rand 3	Store 3 as the seed.
MATH ▶ 3:iPart(30 x MATH ▶ ▶ ▶ 1:rand) + 1 ENTER ENTER ENTER ENTER ENTER ENTER etc.	iPart(6*rand)+1 2 4 5 3 1 2	Get the integer part of the random number that has been multiplied by 6 and add 1. The random numbers are 2, 4, 5, 3, 1, 2, etc. Keep pressing ENTER to get as many numbers as you wish.

METHOD 2

Keystrokes	*Screen Display*	*Comments*
CLEAR		Turn the calculator on and clear the Home Screen.
3 STO▶ MATH ▶ ▶ ▶ 1 :rand ENTER	3→rand 3	Store 3 as the seed.
MATH ▶ ▶ ▶ 5 :randInt(1 , 6 , 60) ENTER	3→rand 3 randInt(1,6,60) {2 4 5 3 1 2 2 ...	The lower limit is 1. The upper limit is 6. The number of random integers we wish is 60. Use the right arrow key to see the rest of the list. The random numbers are 2, 4, 5, etc.

If you press ENTER again you will get the next six random numbers.

Example 3

Use the calculator's random number generator to simulate 20 sums of the numbers on two rolled dice. Use 6 as the seed.

Solution:

Keystrokes	*Screen Display*	*Comments*
CLEAR		Turn the calculator on and clear the Home Screen.
6 STO▶ MATH ▶ ▶ ▶ 1 :rand ENTER	6→rand 6	Store 6 as the seed.
MATH ▶ ▶ ▶ 5 :randInt(1 , 6 , 20) + 5 :randInt(1 , 6 , 20) ENTER	6→rand 6 randInt(1,6,20)+ randInt(1,6,20) {6 5 4 11 6 5 6...	The lower limit is 1. The upper limit is 6. The number of random integers we wish is 20. Use the right arrow key to see the rest of the list. The random sums are 6, 5, 4, etc.

Example 4

Use the calculator's random number generator for the Normal distribution to simulate a random sample of 30 from a distribution with $\mu = 23$ and $\sigma = 2$. Use 5 as the seed.

Solution:

Keystrokes	*Screen Display*	*Comments*
CLEAR		Turn the calculator on and clear the Home Screen.
5 STO▶ MATH ▶ ▶ ▶ 1 :rand ENTER	5→rand 5	Store 5 as the seed.
MATH ▶ 6 :randNorm(23 , 2 , 30) ENTER STAT 4 :ClrList 2nd L1 ENTER 2nd ANS STO▶ 2nd L1 ENTER STAT 1 :Edit	5→rand 5 randNorm(23,2,30) {21.7876311 24.... Ans→L1 {21.7876311 24.... L1 L2 L3 1 21.788 ------ ------ 24.169 25.283 23.791 23.335 23.38 23.505 L1(1)=21.78763110...	The list can be viewed from the screen using the right arrow. Note that the numbers are rounded to 7 decimal places. Here we have stored the numbers in list L1 so they can be viewed more easily. Note that only 3 decimal places are displayed in the list whereas 8 decimal places are displayed at the bottom of the screen.

Example 5

Use the calculator's random number generator for the Binomial distribution to simulate a random sample of 30 from a distribution with $n = 8$ and $p = 0.2$. Use 5 as the seed. Graph the results.

Solution:

Keystrokes	*Screen Display*	*Comments*
CLEAR		Turn the calculator on and clear the Home Screen.
5 STO▶ MATH ▶ ▶ ▶ 1 :rand ENTER	5→rand 5	Store 5 as the seed.

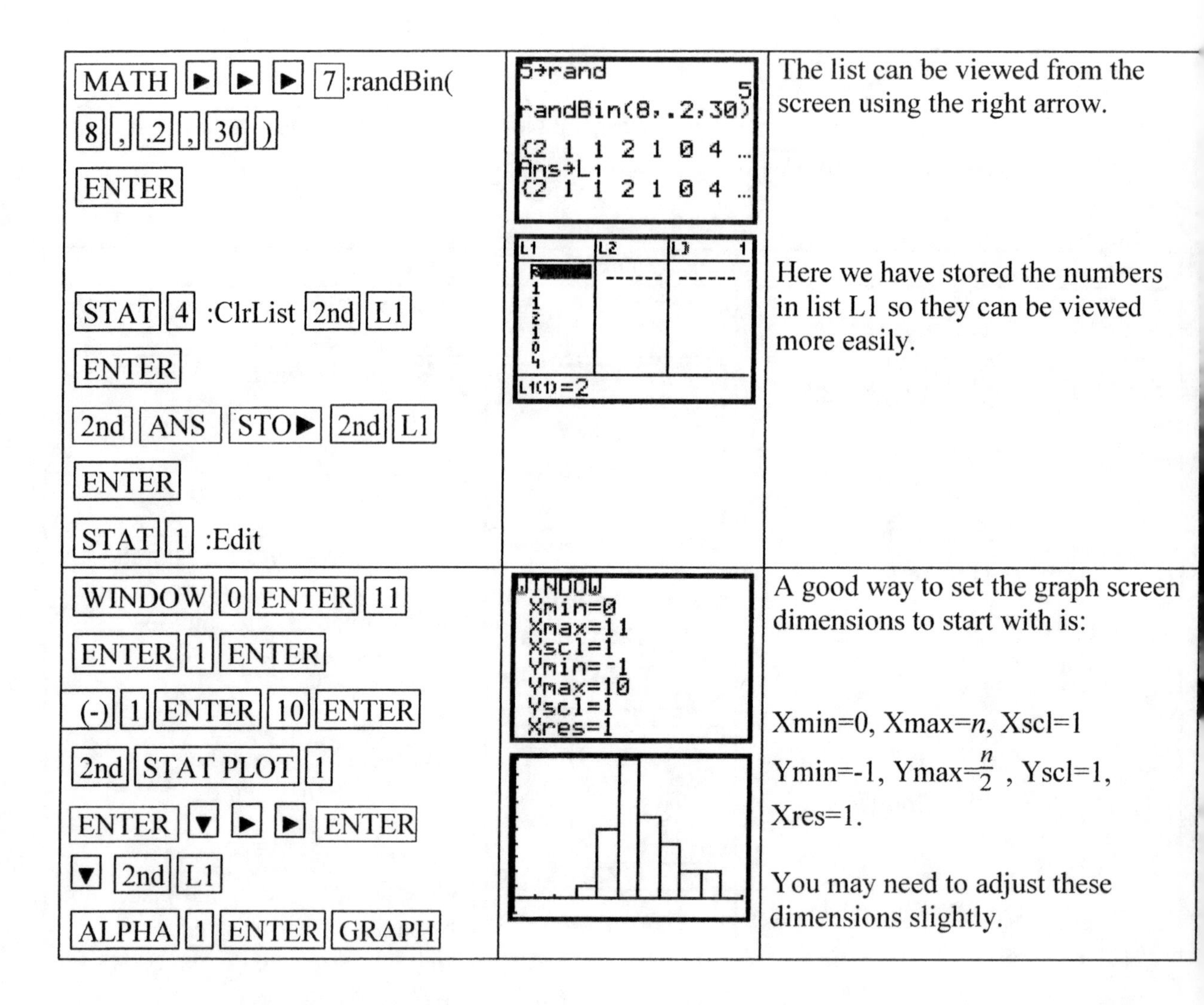

Keystrokes	Screen	Explanation
MATH ▶ ▶ ▶ 7 :randBin(8 , .2 , 30) ENTER	5→rand 5 randBin(8,.2,30) {2 1 1 2 1 0 4 ... Ans→L1 {2 1 1 2 1 0 4 ...	The list can be viewed from the screen using the right arrow.
STAT 4 :ClrList 2nd L1 ENTER 2nd ANS STO▶ 2nd L1 ENTER STAT 1 :Edit	L1 L2 L3 1 / 2 1 1 2 1 0 4 / L1(1)=2	Here we have stored the numbers in list L1 so they can be viewed more easily.
WINDOW 0 ENTER 11 ENTER 1 ENTER (-) 1 ENTER 10 ENTER 2nd STAT PLOT 1 ENTER ▼ ▶ ▶ ENTER ▼ 2nd L1 ALPHA 1 ENTER GRAPH	WINDOW Xmin=0 Xmax=11 Xscl=1 Ymin=-1 Ymax=10 Yscl=1 Xres=1	A good way to set the graph screen dimensions to start with is: Xmin=0, Xmax=n, Xscl=1 Ymin=-1, Ymax=$\frac{n}{2}$, Yscl=1, Xres=1. You may need to adjust these dimensions slightly.

EXERCISE SET 14

1. A twenty-sided die has faces labeled 1 to 20.

   (A) Use the calculator's random number generator to simulate 40 tosses of this die.

   (B) Calculate the percentage of the total number of tosses each of the numbers appears. Are they close to what you expected from a fair die?

   (C) Use the χ^2 Goodness of Fit test to test the hypotheses:

   H_0: The random number generator simulates a fair die.

   H_1: The random number generator does not simulates a fair die.

   (D) Repeat Parts (A) and (B) for 80 tosses. Compare the percentages. Are they closer to what you expect from a fair die?

   (E) Conjecture what will happen to the percentages if 160 tosses are simulated.

2. An eight-sided die has faces labeled 2 to 9.

   (A) Use the calculator's random number generator to simulate 40 tosses of this die.

   (B) Calculate the percentage of the total number of tosses each of the numbers appears.

   (C) Use the χ^2 Goodness of Fit test to test the hypotheses:

   H_0: The random number generator simulates a fair die.

   H_1: The random number generator does not simulates a fair die.

   (D) Repeat Parts (A) and (B) for 80 tosses. Compare the percentages. Are they closer to what you expect from a fair die?

   (E) Conjecture what will happen to the percentages if 120 tosses are simulated.

3. Use the calculator's random number generator to simulate 10 sums of the numbers on one rolled die and a spinner having 4 equal parts. Use 4 as the seed.

4. Use the calculator's random number generator to simulate 20 tosses of two coins. Use the coding 1=Heads and 2=Tails.

5. Use the calculator's random number generator for the Normal distribution to simulate a random sample of 30 from a distribution with $\mu = 3.8$ and $\sigma = 0.2$ Use 5 as the seed.

6. Use the calculator's random number generator for the Normal distribution to simulate a random sample of 30 from a distribution with $\mu = 5382$ and $\sigma = 358$. Use 5 as the seed.

7. Use the calculator's random number generator for the Binomial distribution to simulate a random sample of 30 from a distribution with $n = 10$ and $p = 0.86$. Use 5 as the seed. Graph the results.

8. Use the calculator's random number generator for the Binomial distribution to simulate a random sample of 30 from a distribution with $n = 27$ and $p = 0.4$. Use 5 as the seed. Graph the results.

NOTES

ANSWERS TO SELECTED EXERCISES

Chapter 1

1 (A)

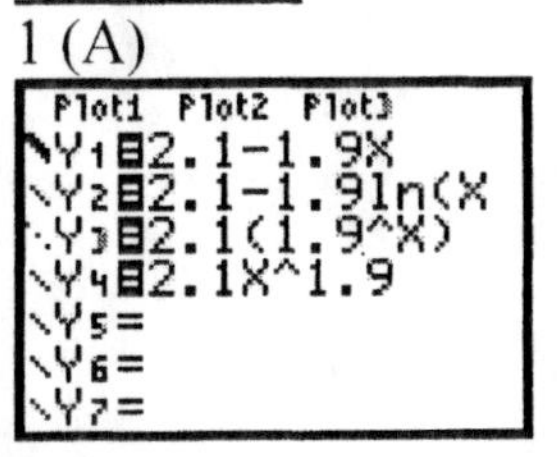

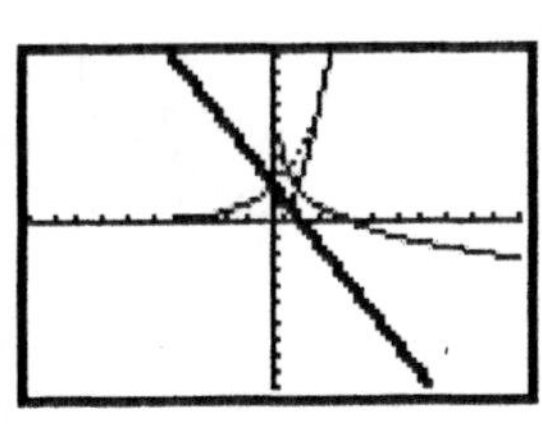

1 (B)

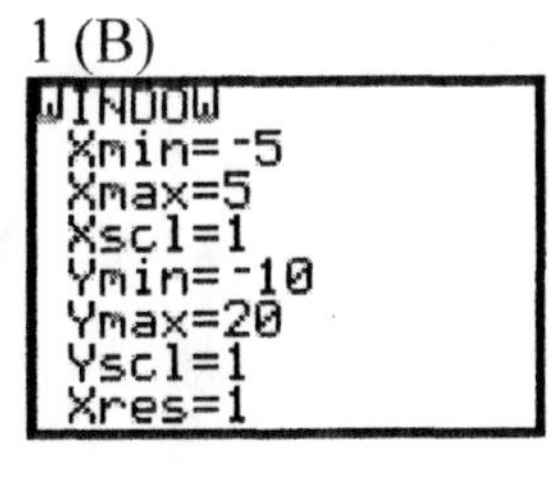

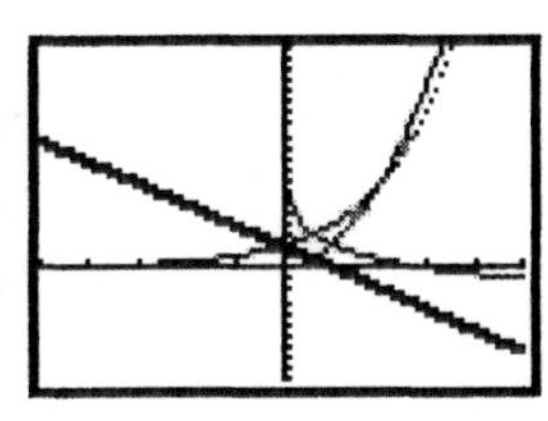

2 (A)

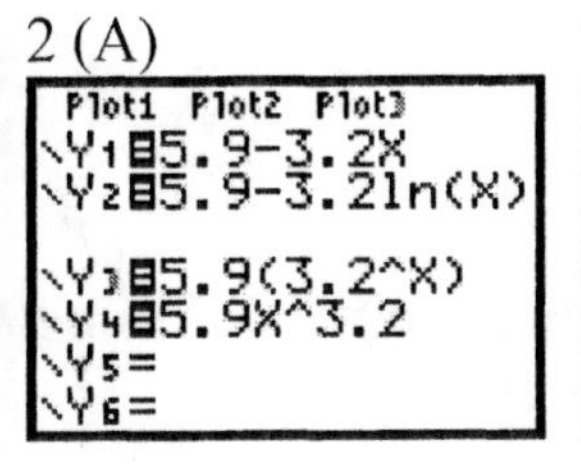

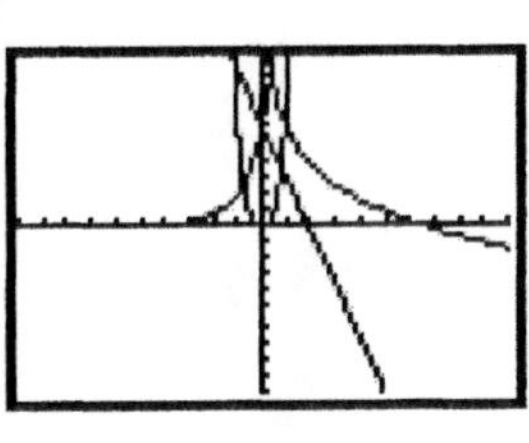

2 (B)

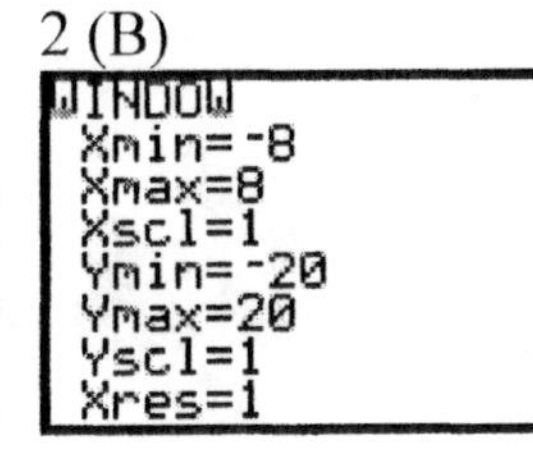

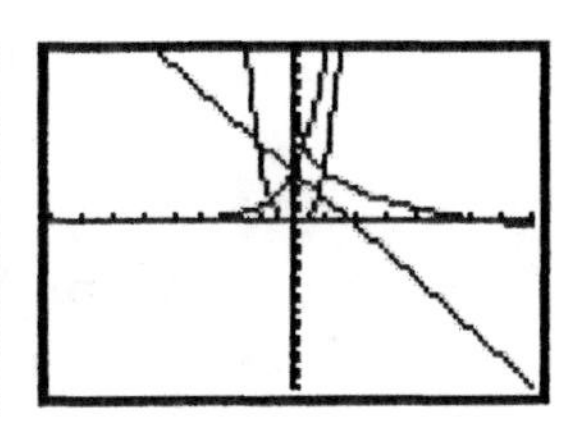

3. $x = 1.08$

4. $x = 0.38$

Chapter 2

1. Set the graph screen dimensions at Xmin=0, Xmax=60, Xscl=10, Ymin=0, Ymax= 20, and Yscl=1. Start the table at 0 and end at 60 so that there is space on the right and left of the histogram. The scale for the x variable is the class width.

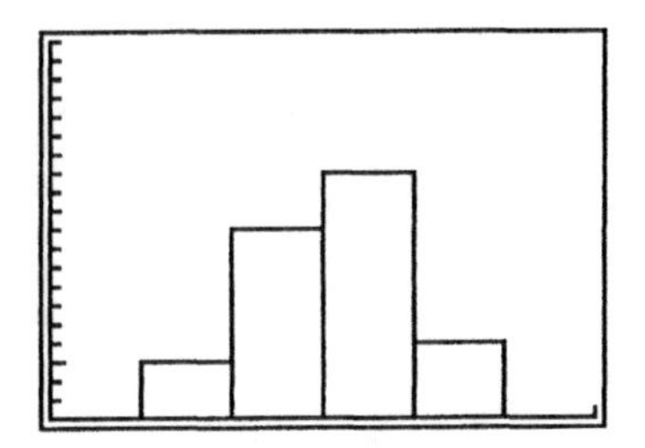

\$0 and under \$10	0
\$10 and under \$20	3
\$20 and under \$30	10
\$30 and under \$40	13
\$40 and under \$50	4
\$50 and under \$60	0

3. Set the graph screen dimensions at Xmin=0, Xmax=10000, Xscl=500, Ymin=0, Ymax= 20, and Yscl=1.

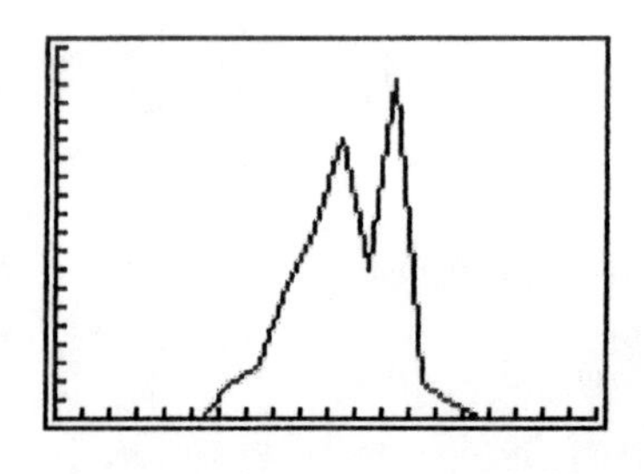

Class Midpoints	Number of Cars (Frequencies)
2750	0
3250	2
3750	3
4250	7
4750	10
5250	15
5750	8
6250	18
6750	2
7250	0

Include midpoints of 2750 and 7250 so that the graph goes down to the x axis.

5. Set the graph screen dimensions at Xmin=2500, Xmax=7500, Xscl=500, Ymin=0, Ymax= 70, and Yscl=10.

Find the cumulative frequencies.

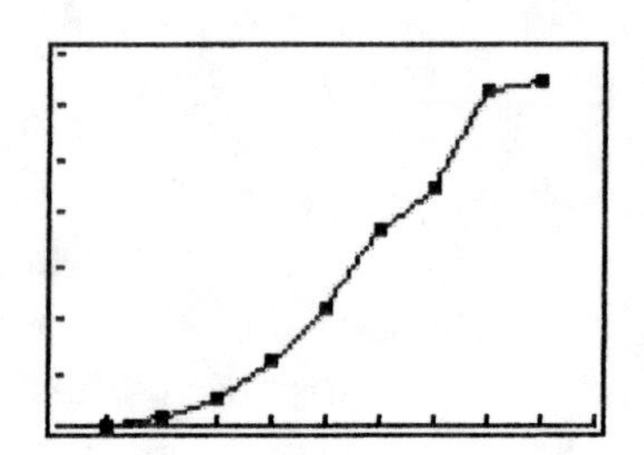

Miles Traveled	Number of Cars (Frequencies)
Less than 3000	0
Less than 3500	2
Less than 4000	5
Less than 4500	12
Less than 5000	22
Less than 5500	37
Less than 6000	45
Less than 6500	63
Less than 7000	65

7. The rightmost two dots are moved downwards.

9. Set the graph screen dimensions at Xmin=0, Xmax=8, Xscl=1, Ymin=50, Ymax= 110, and Yscl=10.

Number the months 1 through 7.

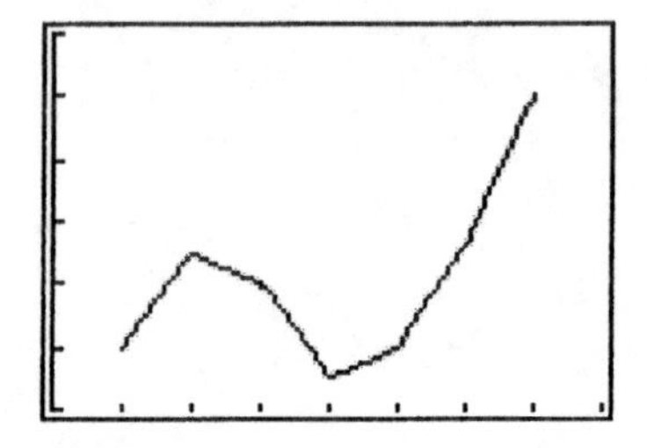

11. Set the graph screen dimensions at Xmin=0, Xmax=15, Xscl=1, Ymin=0, Ymax= 110, and Yscl=10.

    Since 13 is the number assigned to July, use 15 as the maximum x value to allow space on the right of the bar.

    Use odd numbers to represent the months.

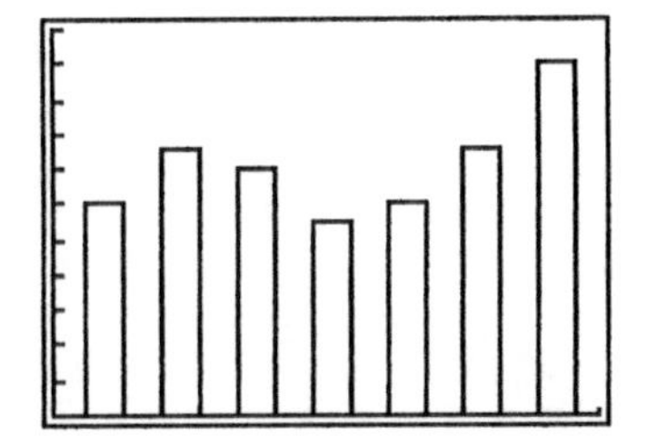

13. (A) Calculate the percents for each category. Set the graph screen dimensions at Xmin=0, Xmax=15, Xscl=1, Ymin=0, Ymax= 0.25, and Yscl=0.01.

    (B) Removing July causes all of the other months' percentages to increase.

Chapter 3

1. (A) 77.7 (B) 8.5 (C) 72.5 (D) 75.0
   (E) 84.0 (F) 77.5 (G) 72.3 (H) 10.9%
   (I) Set graph screen dimensions as Xmin=55, Xmax=100, Xscl=5, Ymin=0, Ymax=3, Yscl=1.

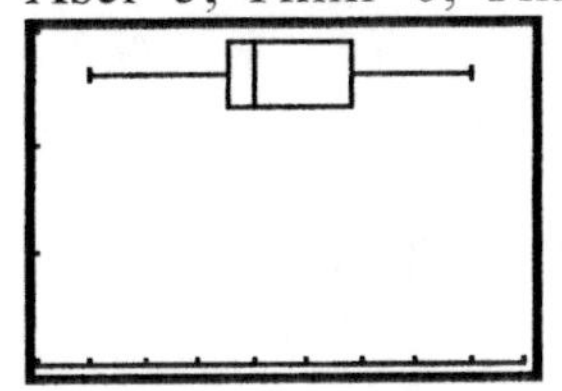

3. (A) 78.3 (B) 7.5 (C) 72.5 (D) 75.0
   (E) 84.0 (F) 82.5 (G) 56.1 (H) 9.6%

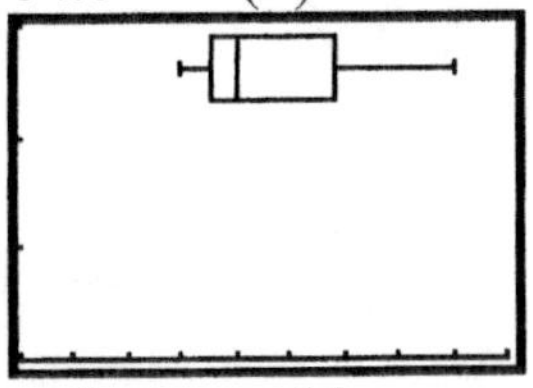

   The mean increased because a data value was increased.

   Q1, Median, Q3 do not change since changing 60 to 70 did not change the order of the data.

   The standard deviation, midrange and the variance also decreased because the spread of the data is decreased. The boxplot left whisker is shorter since 70 is greater than 60.

5. (A) 5319.2 (B) 870.0 (C) 756850.96

7. (A) 5188.6 (B) 850.8 (C) 723840.85

   Effects of the change in the data:

   The mean decreased since we eliminated some data having larger values.

   The standard deviation and variance decreased since some of the large values were eliminated from the data set.

Chapter 4

1. 360 3. 116280 5. 1140 7. 0.0036 9. 0.0829

11. 0.0066 13. (A) 0.0300 (B) 1 - 0.0930 = 0.9070

Chapter 5

1. 1.02 3. (A) 0.4970, 0.3685, 0.1139, 0.0188, 0.0017, 0.0001, 0.0000

(B)

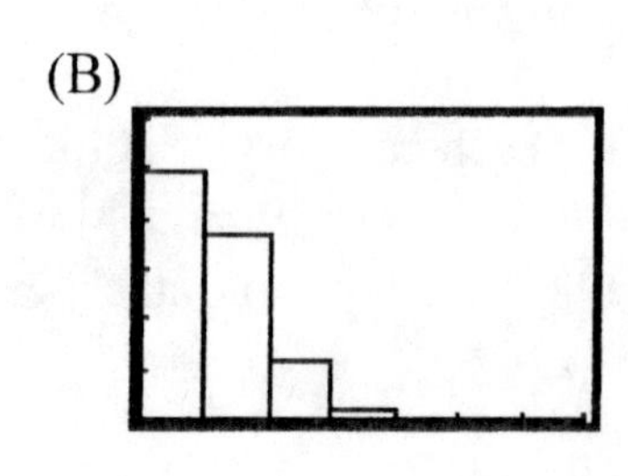

(C) $\mu = 0.66$, $\sigma = 0.77$

(D) 0.1139 (E) 0.9794

5. (A) 0.1954 (B) 1 – *P*(at most 1) = 0.0984

Chapter 6

1. Using ShadeNorm(100, 275, 182, 15)

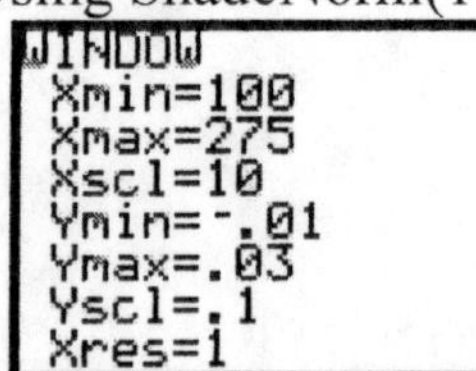

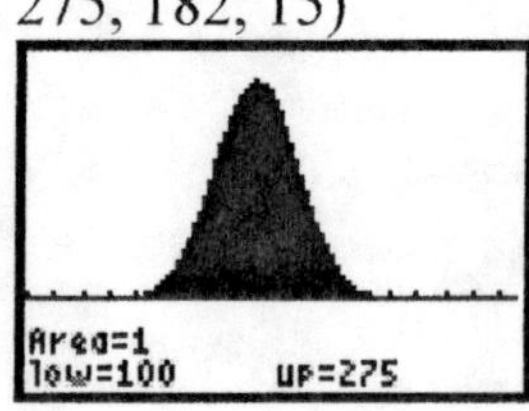

Using the same graph screen dimensions and Y1=normpdf(X,182,15)

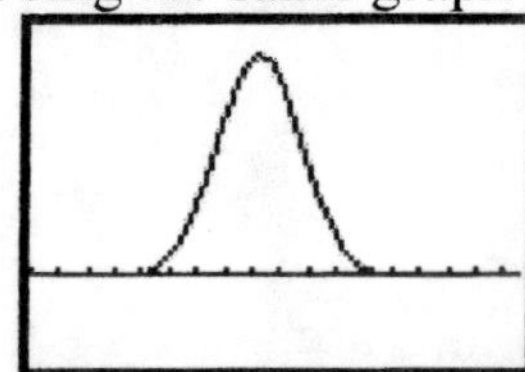

Be sure to clear the drawings first.

3. (A) 0.9645 (B) 0.7734 (C) 71.4500

5.

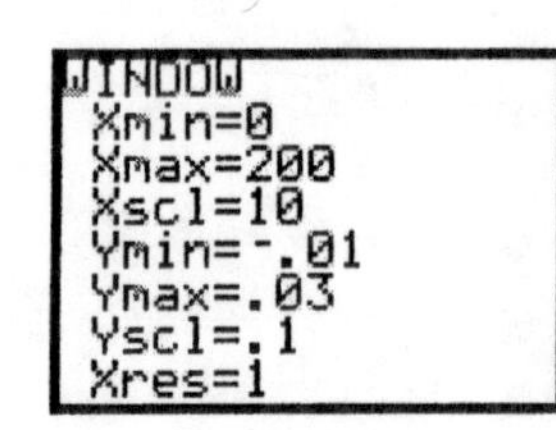

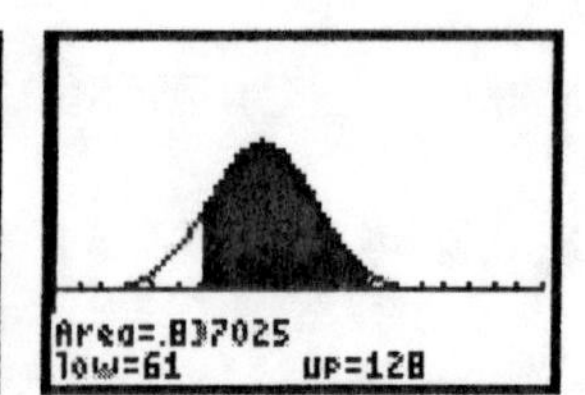

7. (A) 2.3553 (B) 2.3582 (C) 2.3641 (D) It approaches 2.37.

Chapter 7

1. 46.9 to 48.7   3. 2.5 to 2.7   5. 0.767 to 0.933

7. [2nd] [DISTR] [3] :invNorm([.975] [)] [ENTER] [x] [80] [÷] [25] [ENTER] [x^2] [ENTER]
   At least 40.

Chapter 8

1. (A) The critical value is -3.38.

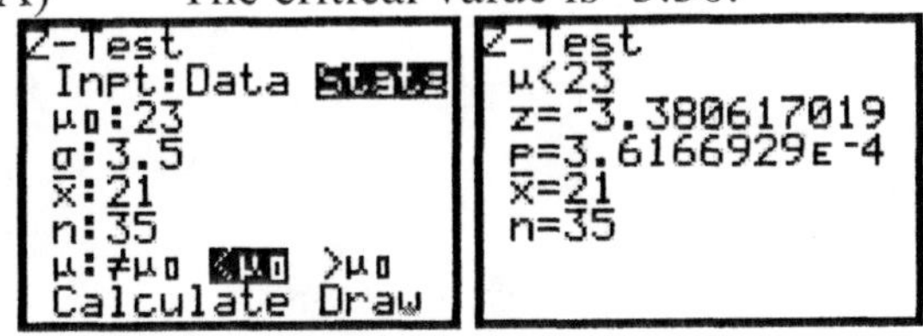

   (B) Change n to 70. Now the critical value is −4.78
   (C) The critical value decreases as the sample size increases. No it does not double in size.

3. (A) Yes because 0.0004 < 0.01
   (B) Yes because 0.00000087 < 0.01.
   (C) No because both *P*-Values are less than the level of significance.

5. No, since the critical value is now 1.609 with a *P*-Value of 0.054.

7. This is a two-tailed test. The critical value is 4.16 and the *P*-Value is 0.0016. The theory is refutable.

9. No. *P*-Value = 0.31

11. The test statistic is 35.097. The *P*-Value is 0.1017.

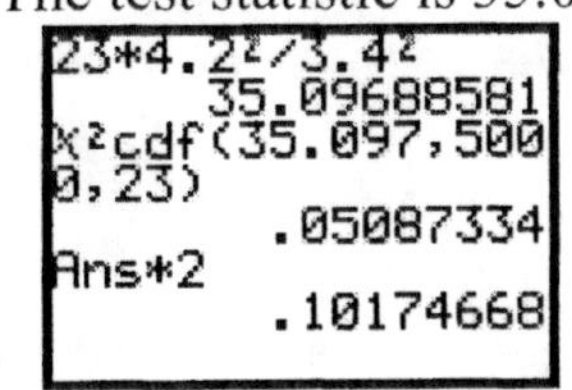

13. (A)

ZInterval
(19.476,22.524)
x̄=21
n=35

(B)

ZInterval
(19.922,22.078)
x̄=21
n=70

Both 99% confidence intervals on the population mean do not contain the hypothesized value of 23 lbs. The decision in Problems 3(A) and 3(B) was to reject the null hypothesis. These results agree.

15. (A)

WINDOW
Xmin=0
Xmax=30
Xscl=1
Ymin=-.05
Ymax=.15
Yscl=.1
Xres=1

(B)

WINDOW
Xmin=0
Xmax=40
Xscl=1
Ymin=-.05
Ymax=.1
Yscl=.1
Xres=1

Area=.139275 df=18
low=7.52 up=12.03

Chapter 9

1. (A) The *P*-Value is 0.0002. Reject the null hypothesis

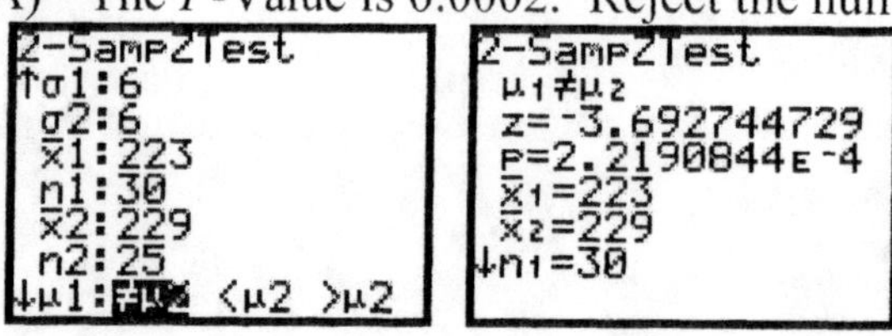

(B) The interval has negative values since sample 1 has a lower sample mean than sample 2. The interval is -10.19 to -1.82.
To get positive values, interchange sample 1 with sample 2.

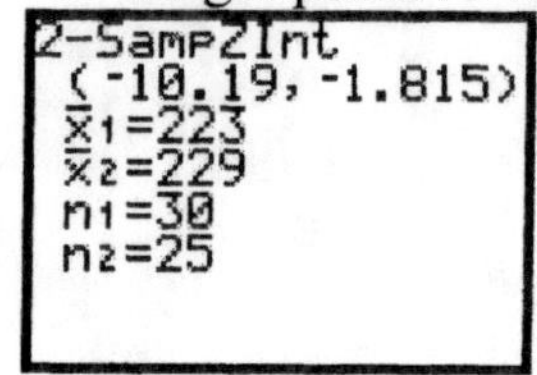

3. (A) The *P*-Value is 0.28. Do not reject the null hypothesis. There is not enough evidence to conclude that the single drivers do more driving for pleasure trips on average than married drivers.

(B) The confidence interval is -8.10 to 12.67. However, we see the standard deviations are almost the same value. Using pooled variances the confidence interval is the same when rounded to two decimal places.

Not Pooled   Pooled

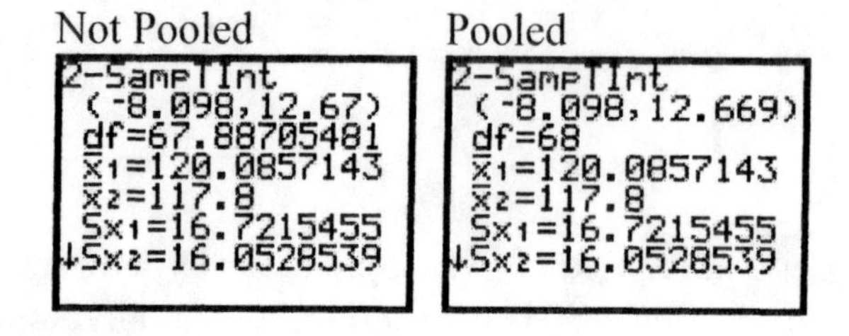

5. Clear the drawings first by using [2nd] [DRAW] [1] :ClrDraw [ENTER] .

   Change the graph screen dimensions to Xmin=0, Xmax=5, Xscl=1, Ymin= -0.1, Ymax=1, Yscl=1, Xres=1

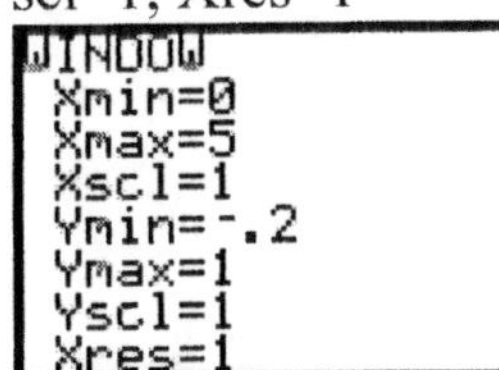

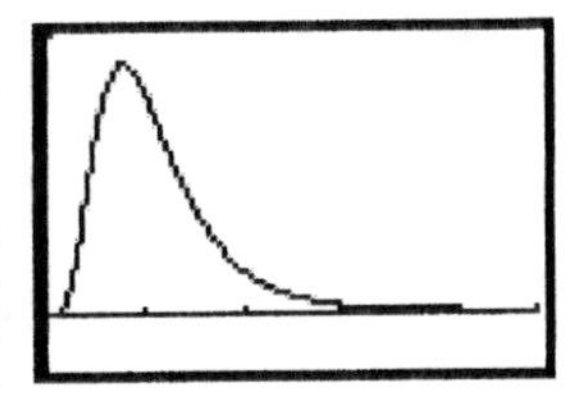

7. Clear the drawings first by using [2nd] [DRAW] [1] :ClrDraw [ENTER] .

   Change the graph screen dimensions to Xmin=0, Xmax=5, Xscl=1, Ymin= -0.3, Ymax=1, Yscl=1, Xres=1

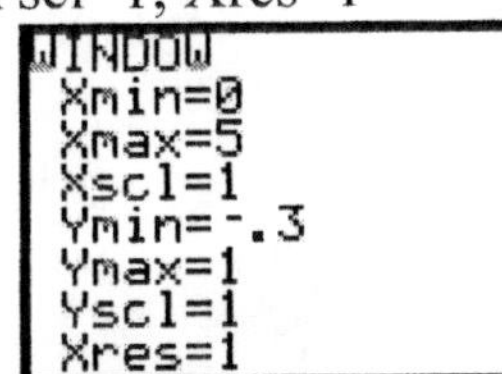

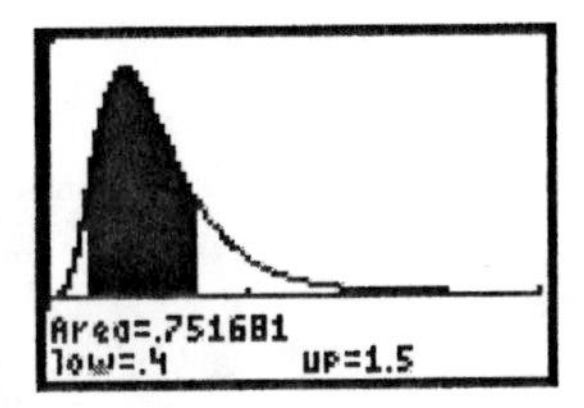

9. The *P*-Value is 0.77. Do not reject the null hypothesis. There is not enough evidence to conclude that the amount spent per pupil in the two states is different.

```
2-SampFTest
 σ1≠σ2
 F=1.135407745
 p=.7684082702
 Sx1=28.12
 Sx2=26.39
↓n1=20
```

11. Enter the data into list L1 and L2. Calculate the difference and place in L3 using [L2] [-] [L1] [STO ▶] [L3]. Use the one-sample *t* test on L3.

    The *P*-Value is less than the level of significance. Reject the null hypothesis. There is enough evidence to conclude that the tutoring sessions helped to improve the students' vocabulary.

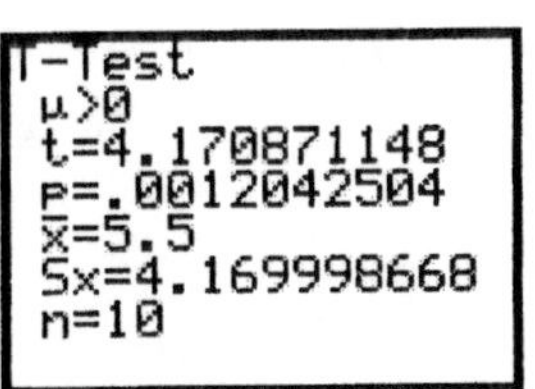

13. This problem has whole numbers for the number of items. If a problem has a percentage given of the total, this percentage must be converted to a whole number. If the division results in a decimal portion, then multiply both the result and the total number by a power of 10 until a whole number results. The calculator 2-PropZTest will only calculate using whole numbers for x_1 and x_2.

(A) The *P*-Value for this test is 0.35 which is greater than the level of significance. Do not reject the null hypothesis. There is not enough evidence to conclude that the proportion of nonreaders in the two cities is difference.

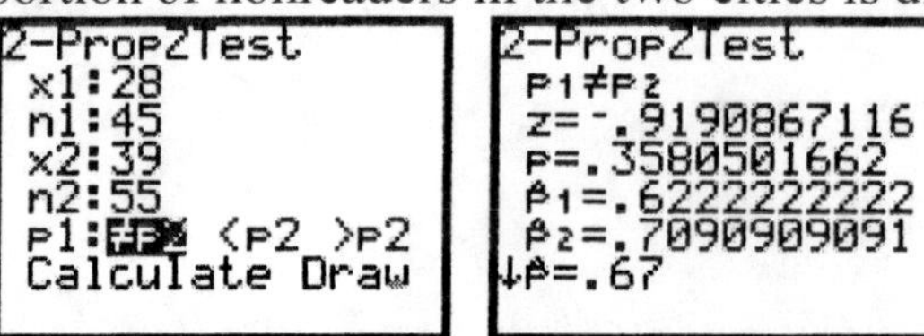

(B) The confidence interval is -0.33 to 0.16.

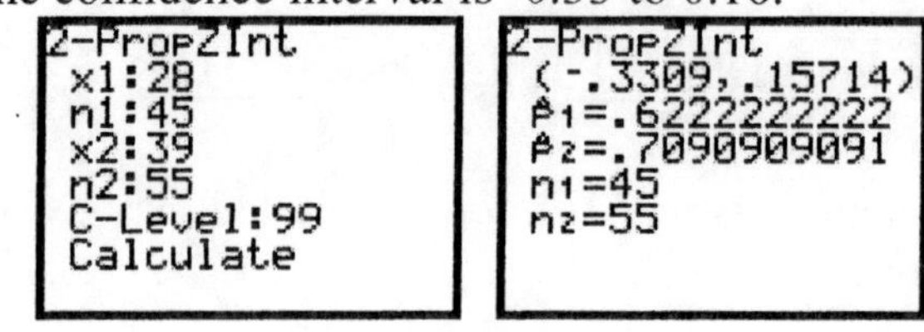

Chapter 10

1. (A)

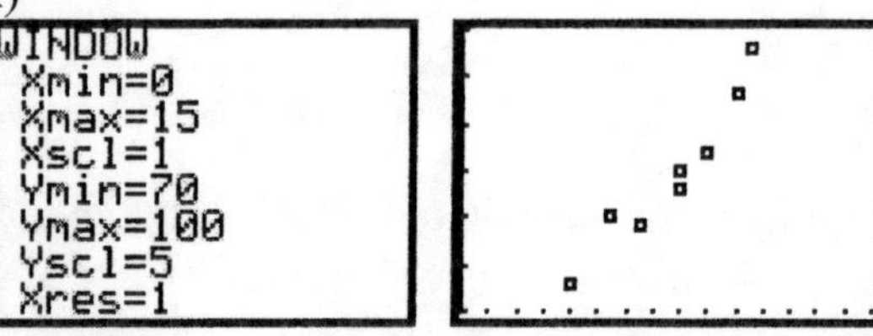

(B) & (F) The linear correlation coefficient is 0.96. The coefficient of determination is 0.91. This means that 91.5% of the variation in the dependent variable can be explained by the independent variable and the regression.

(C) The regression equation is $y' = 3.41 + 58.50x$

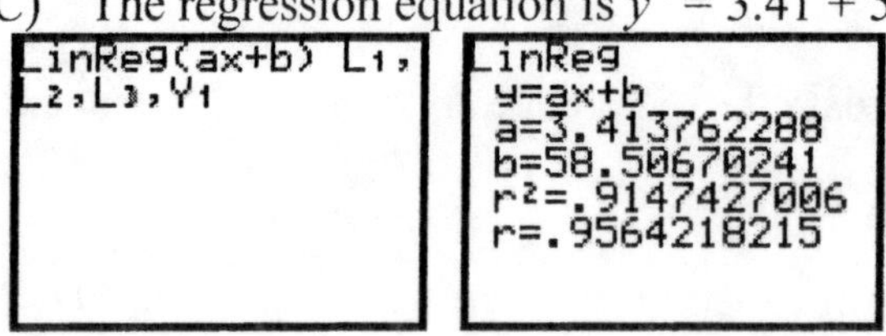

(D)

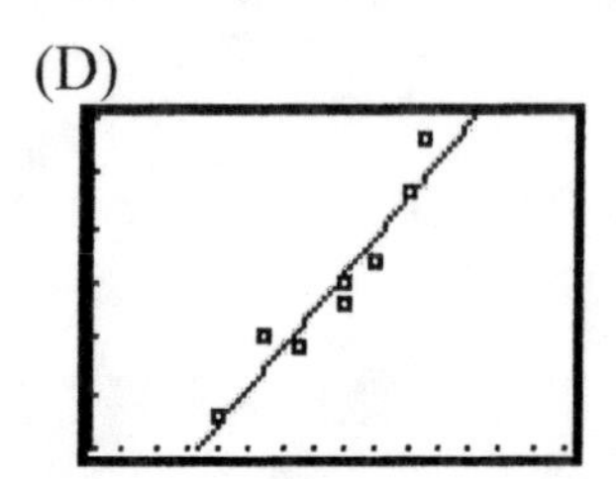

(E) The *P*-Value of the hypothesis test is 0.0001. Reject the null hypothesis. There is enough evidence at the 0.05 level of significance to conclude that there is positive correlation between the variables.

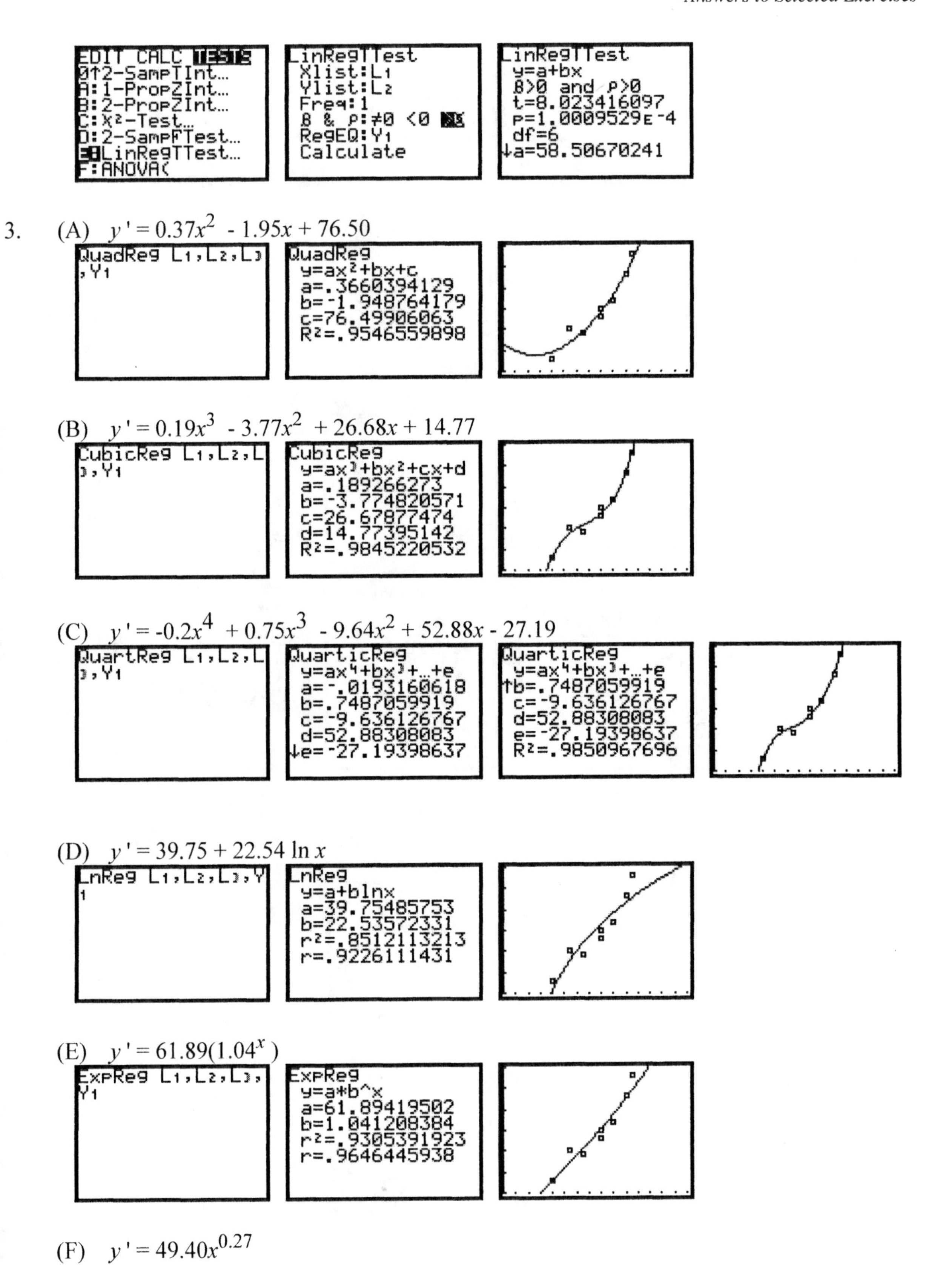

3. (A) $y' = 0.37x^2 - 1.95x + 76.50$

(B) $y' = 0.19x^3 - 3.77x^2 + 26.68x + 14.77$

(C) $y' = -0.2x^4 + 0.75x^3 - 9.64x^2 + 52.88x - 27.19$

(D) $y' = 39.75 + 22.54 \ln x$

(E) $y' = 61.89(1.04^x)$

(F) $y' = 49.40x^{0.27}$

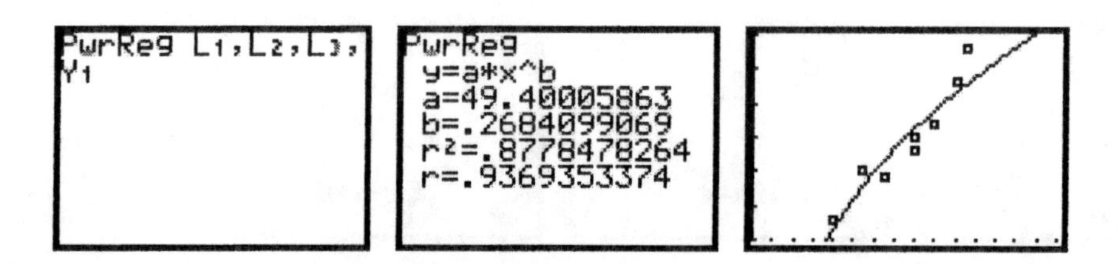

Chapter 11

1. (A) Expected value for each day is 27.
   Test value is 17.77.

   (B) Set the graph screen dimensions as shown below.
   Select the Shadeχ^2 option from the DISTR DRAW menu.

   WINDOW
   Xmin=0
   Xmax=30
   Xscl=1
   Ymin=-.05
   Ymax=.15
   Yscl=.1
   Xres=1

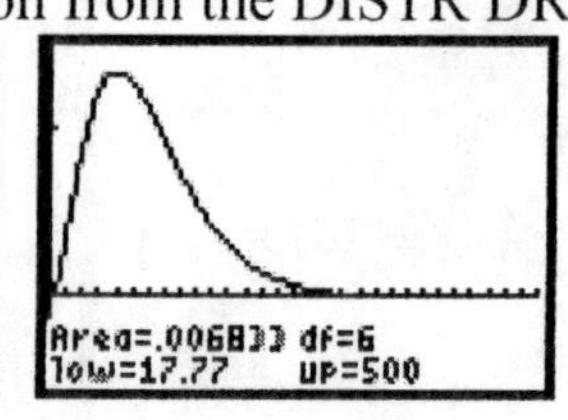

   (C) Set the graph screen dimensions as shown below.
   Set Plot1 and Plot2 as shown below.

   WINDOW
   Xmin=0
   Xmax=8
   Xscl=1
   Ymin=0
   Ymax=45
   Yscl=1
   Xres=1

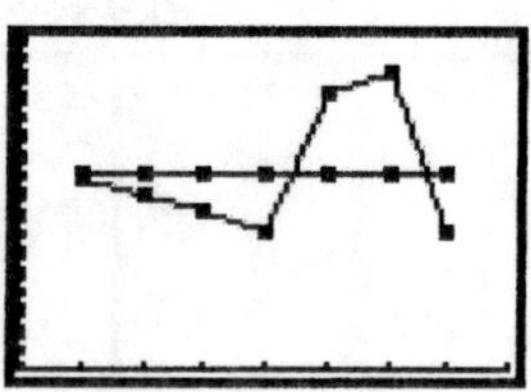

3. (A) The *P*-Value is not less than 0.05. Do not reject the null hypothesis.
   There is not enough evidence to conclude that the age of an individual is related to coffee consumption.

   MATRIX[A] 4 ×3

18	16	12
9	15	27
8	12	10
13	10	9

   (B) The graph screen dimensions are automatically set using the Draw option.

   χ²-Test
   Observed: [A]
   Expected: [B]
   Calculate Draw

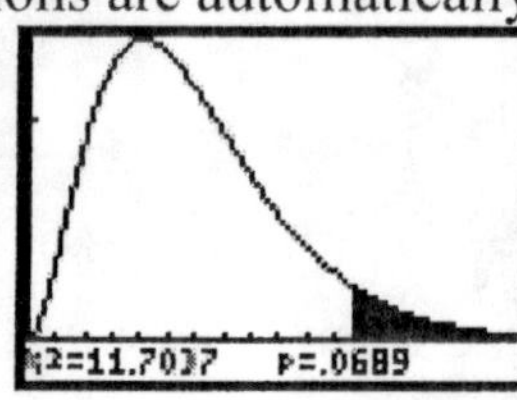

Chapter 12

1. The *P*-Value is 0.03. Reject the null hypothesis. There is enough evidence to conclude that there is a difference in the average times employees travel to work.

L1	L2	L3
35	9	15
18	3	6
27	12	27
24	6	42
------	14	18
	8	
	21	

L3(6) =

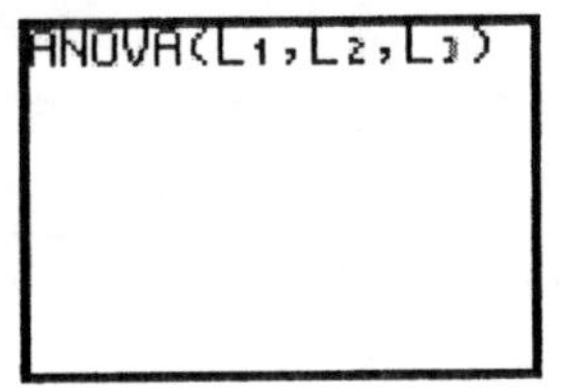

One-way ANOVA
F=4.244583452
p=.0381245325
Factor
df=2
SS=721.523214
MS=360.761607

One-way ANOVA
MS=360.761607
Error
df=13
SS=1104.91429
MS=84.9934066
Sxp=9.21918687

3. Use Xmin=0, Xmax=4, Xscl=1, Ymin=130, Ymax=200, Yscl=5

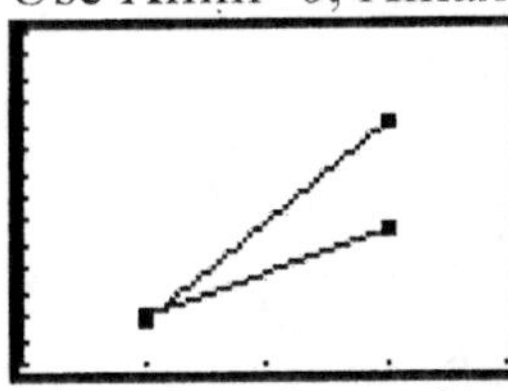

Changing the scale to Xmin=0.8, Xmax=1.2, Xscl=0.1, Ymin=139, Ymax=141, Yscl=0.5
We see that the lines cross indicating an interaction.

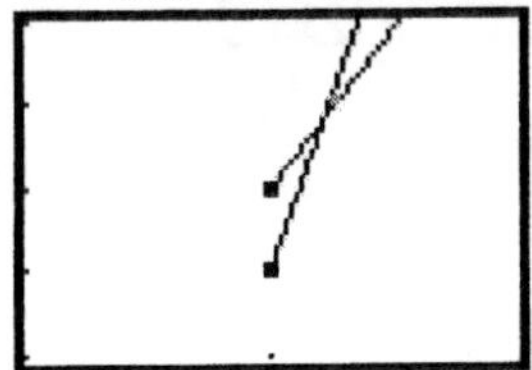

Chapter 13

1. (A) There are 3 data values less than $325. There are 9 data values greater than $325. The test value is 3.

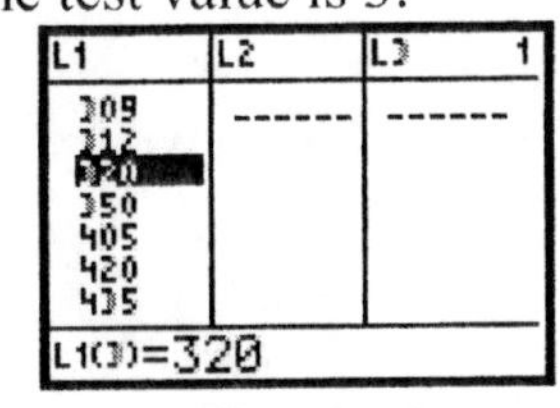

L1	L2	L3
460		
450		
435		
420		
405		
350		
320		

L1(9)=350

(B) The sum of the ranks for negative values is 5.
The sum of the ranks for positive values is 31.
Hence, the test value is 5.

3. There is 1 negative value. There are seven positive values. The test value is 1.

L1	L2	L3
187	178	
163	62	
201	188	
158	156	
139	133	
143	150	
198	175	

L3(1)=

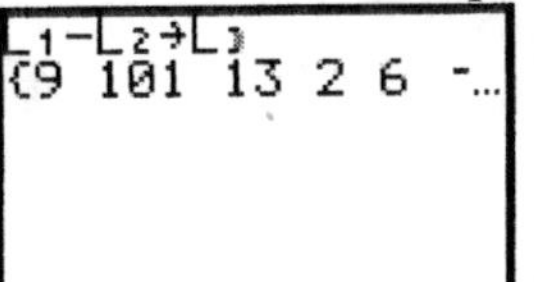

L1	L2	L3
143	150	-7
158	156	2
154	150	4
139	133	6
187	178	9
201	188	13
198	175	23

L3(1)= -7

5. Enter the data. All data screens are shown.

```
L1    L2    L3   2 | L1   L2   L3   2 | L1   L2   L3   2 | L1   L2   L3   2
102   1     ------ | 91   2         | 60   3         | 62   4
108   1            | 93   2         | 56   4         | 103  5
99    1            | 82   2         | 43   4         | 98   5
97    1            | 74   3         | 52   4         | 94   5
92    1            | 83   3         | 58   4         | 89   5
87    2            | 78   3         | 62   4         | 88   5
86    2            | 74   3         | 103  5         | ------
L2(7) =2           | L2(14) =3      | L2(21) =5      | L2(26) =
```

Sort on L1.

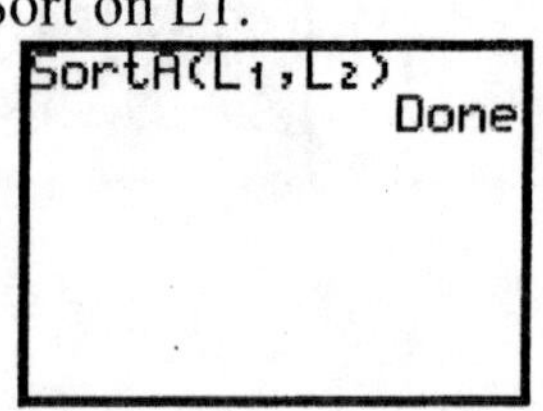

Assign Ranks. First and last screens are shown below.

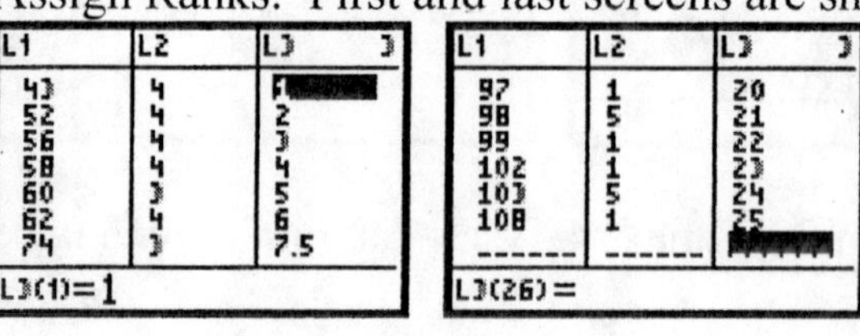

Sort on L2.

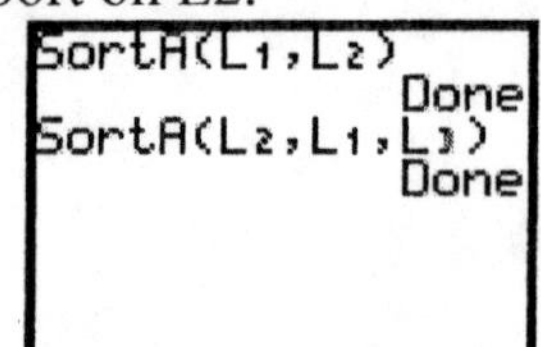

Add ranks in each group.

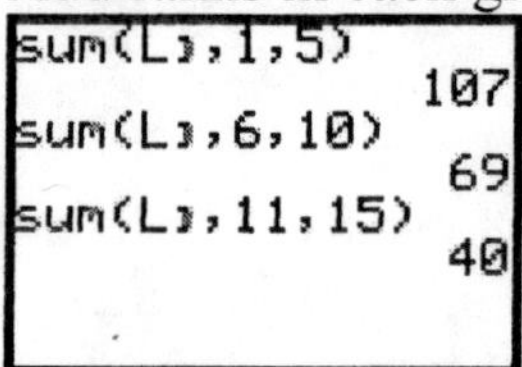


```
                 69
sum(L3,11,15)
                 40
sum(L3,16,20)
                 16
sum(L3,21,25)
                 93
```

Calculate the test value. The quantity to calculate is:

$$H = \frac{12}{N(N+1)}\left(\frac{R_1^2}{n_1}+\frac{R_2^2}{n_2}+\frac{R_3^2}{n_3}+\frac{R_4^2}{n_4}+\frac{R_5^2}{n_5}\right) - 3(N+1)$$

where $R_1 = 107$, $R_2 = 69$, $R_3 = 40$, $R_4 = 16$, and $R_6 = 93$, $N = 25$.
The test value is 168.42.

7. $\Sigma d^2 = 42$. $r_s = 0.5$

Chapter 14

1. Answers may vary. (C) The expected number for each side with $n = 40$ is 2 or 5%.

3. The sums are 7, 7, 8, 8, 6, 7, 5, 4, 5, and 3.

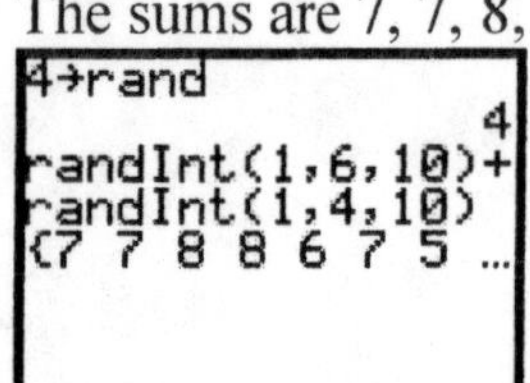

5.

7.

```
5→rand
                 5
randBin(10,.86,3
0)
{10 9 8 7 10 8 ...
```